More Praise for *Billion Dollar Apps*

"Could he make it any easier? Every organization, regardless of function, needs a mobile strategy today; yet most don't know where to start. Alex's 6 Step approach gives you the roadmap and tools to plan and execute your mobile platform along with the appropriate metrics to measure your success."

JANET VIANE
Vice President, Marketing Operations
Sears Holdings Corporation

"We all know someone who 'has a son or in-law who can develop an app that solves all of our business problems.' Deploying apps and devices should be as stress-free as Apple and Google make them seem. But the real world is seldom as cut and dry. Alex helps take the guess work out of analyzing, developing and delivering on the mobile world's billion dollar potential that we are all chasing. I wish I had Alex's BDA book before I started my deployments; the clear examples work through the process and aid practitioners in identifying common pitfalls."

JON MERRITT
Senior Manager, Flight Operations Technology
United Airlines

BILLION
DOLLAR APPS

How to Find and Implement
a Winning Mobile Strategy

Alex Bratton

Innovation Networks
Chicago

Copyright © 2014 Alex Bratton

All rights reserved. No part of this book may be reproduced or transmitted in any form or by any means, electronic or mechanical, including photocopying, recording, or by any information storage and retrieval system without the prior written permission of the author, except for the inclusion of brief quotations in critical reviews and certain other noncommercial uses permitted by copyright law. For permission requests, contact alexb@lextech.com.

Library of Congress Cataloging-in-Publication Data is available for this title.

LCCN: 2013923416
ISBN: 978-0985771096
e-ISBN: 9780985771089

ACKNOWLEDGEMENTS

Bringing *Billion Dollar Apps* to the world has taken a year of research, interviews and production. I really appreciate the assistance of a bunch of great people in making it happen.

I'd like to thank Ruairi Breathnach, Kevin Glynn, Tony Dillon, Andy Graham, Neil Mylet, Robert Sarkis, and Matt Hartzman for their insights and sharing their real world stories.

Brian Bernardi and Dao Yang put together some awesome illustrations and cover art that really bring the concepts to life.

Thanks to Adrienne Szewczyk, Mark Sherbin, and Thomas Caprel, Jr. - your dedication to the production process has made this book a reality. Thank you for bringing sanity to the project on everything from project organization to layout to digital publishing magic.

Will Scott, my partner at Lextech Global Services, understood the vision for why this message needed to be shared and took on even more of the business so I could focus on the book.

Thanks to Vistage 363 members and those mobile experts at companies I can't name publicly for your feedback and ideas (you know who you are).

None of this would have been possible without my wife and Muse, Michelle. Her support and encouragement have kept me going all these years.

This book is dedicated to the memory of my mother-in-law, Sherri, whose life shift from no tech to passionate mobile user was inspiring.

CONTENTS

1. Why Go Mobile? — 1
2. How Mobile Changed The World — 23
3. How Protectus Discovered Billions of Potential in One Year — 37
4. What Is A Mobile Change Agent? — 71
5. Billion Dollar Apps Process Overview — 91
6. Step One: Goals in Perspective — 97
7. Step Two: Finding the Workflows that Form the Process — 101
8. Step Three: Identifying Workflow Issues — 113
9. Step Four: Generating App Ideas — 125
10. Step Five: Calculating the Impact of Your Concepts — 131
11. Step Six: Scoring and Prioritizing Your Concepts — 137
12. Considerations for App Development and Launch — 143
13. Taking the Plunge — 177

CHAPTER 1
Why Go Mobile?

Billions of dollars lie untouched by your industry, hidden by inefficiencies that have crept into workflows over the years. With the right technology, your company has the opportunity to break through the barriers that stop it from claiming that revenue. You can be the one to make it happen.

In today's mobile society, businesses can't afford to miss out on that billion-dollar opportunity. To attain it, you'll need a progressive approach, a bit of knowledge, and the tools for the job.

Most important, understand that tablets and smartphones are means, not ends. In the long run, powerful change comes from your ability to see the dozens of opportunities for growth that mobility can address across your organization.

How do you make mobility practical? What's your business missing that dozens of Fortune 500 companies already have? Where do you even start? The answer is in the apps.

I wrote this book because I can't stand to see more half-baked, crappy apps that don't deliver on the true potential of mobility. We need to fix this problem.

Mobile business is about more than equipping staff with smartphones. It's about making processes easier for people. It's about trimming the fat from workflows.

Mobile business takes shortcuts that make your processes *better, faster, and more lucrative.*

Think of your current processes and workflows as a convoluted map of twisted pathways that overlap and dead-end. By the time you reach your goal, you've started and stopped so often that it's taken you ten times longer than if you'd walked a straight line.

But mobility itself is only a supporting theme of this book. Apps are your toolset. What I hope to impart to you is much bigger than either of these concepts. I want you to understand that Billion Dollar Apps revolutionize your business processes.

Without the ability to take a critical and necessary look at those processes...well, your organization could stay stuck in a rut that might *cost* you billions over time. The stakes are high. The process takes some effort, but it's very straightforward— with the help of this book.

Billion Dollar Apps will require you to step entirely outside of your traditional perspective on the workflows that make up your organization. You must prepare yourself to see workflows at their most basic levels and make changes that can radically alter your approach to business.

One ambitious entrepreneur did exactly that. What's most remarkable about his project is that he started from scratch, with no technology to form a foundation for the major shift in the process. Here's his story:

HOW TO MOBILE-ENABLE A GRAIN SILO

In 2010, Neil Mylet brought mobility to a grain silo. Going mobile halved the labor he needed to complete a crucial business process. It made the process 10 to 20 percent more effective. It also eliminated health hazards that once plagued workers.

It all started when the Indiana entrepreneur cast a critical eye on the loading process for grain-hauling trucks. Loading up the trucks was a messy (and sometimes dangerous) process for the truck drivers. Neil knew there was a way to make things better, but he wasn't sure what it was. He asked the key question: Can we give the driver visibility and control of the processing using an iPhone?

So, he approached my company, Lextech. We sat down together and analyzed the typical workflows for a truck driver. To start, the driver pulls his hauling truck beneath the spout of a grain elevator, a pipe that drops grain into the vehicle's large storage area. He opens the door, turns around, and stands on the back of the cab. His new position helps him confirm that the trailer is aligned with the spout.

The driver signals approval to the ground operator, who pulls a chain. Industrial motors, augers, and other equipment kick into action, pumping grain into the truck's trailer. When the pile in the front part of the trailer gets big enough, the driver slips back into the cab, pulling the truck forward a few feet so grain can fill the next part of the trailer. Then he's back out of the cab to watch that section fill.

When the trailer is full, the person on the ground pulls the chain to stop everything, and the truck drives off. There are no laptops and no web apps here. It's purely a no-tech workflow.

Neil saw two major problems with the process.

It's inefficient.

Grain loading required two people. Both workers stand around doing nothing for the majority of the time. Theoretically, one person could finish the job if the switch could be flipped remotely.

It's dangerous.

Even the most dexterous truck drivers had lost their balance and fallen off cabs while twisting around to monitor pile sizes. And that's not all. Grain kicked up dust into the driver's face as it fell. Inhaling large amounts of it can cause serious lung problems.

Together, we found a way to eliminate both issues. All it took was the right question: How could the driver control these workflows from inside the cab? The answer required a new approach, the necessary hardware, and the right apps.

Here's how it works:

As the truck pulls beneath the grain spout, the driver's iPhone app automatically connects to control equipment, a Wi-Fi–enabled yellow box mounted to the side of the silo.

- The driver pushes a button on the iPhone that activates the machinery. Grain starts flowing through the equipment, and the spout dumps grain into the trailer.

- A camera above the action transmits video to the driver's iPhone and he watches as the trailer fills. The driver can

- move forward slowly to accommodate more grain at the right time.
- When the trailer is full, the driver pushes the button to deactivate the machinery. Any malfunctions throughout the process trigger an automatic system shutdown.

With the new process, loading grain only requires a single protected worker, eliminating health concerns and reducing labor costs. Somewhere between 10 and 20 percent more trucks can get through the process in a single day, increasing output and revenue potential.

Most importantly, the driver has full visibility and control of the environment from inside the cab. The app empowers the driver with more information and direct control of a previously no-tech workflow. Imagine such efficiency industry-wide. An app truly earns its "billion-dollar" designation.

TRANSFORMING YOUR APPROACH TO BUSINESS PROBLEMS

Neil Mylet is a change agent; He broke with the traditions of his industry. He found new ways to tackle old problems. And he did it by reimagining the solution.

At its core, this is what Billion Dollar Apps (BDAs) are all about: new solutions to old problems. Technology, paired with the right process focus, is what makes it all possible.

Grain loading is a decades-old business process. A trailer full of grain is still the outcome, both before and after Neil's ingenuity. He just found a better way to get there. The new approach saves time and money, frees up resources to focus on other tasks, decreases health risks to workers, and increases revenue potential— all from improving a single process.

This wouldn't be possible without the right mix of technology and innovative thinking. BDAs are about improving business processes with mobility. These two factors must work together for billion-dollar results over the long term. Innovation's biggest impact comes through isolating, deconstructing, and redefining individual workflows.

WHAT ARE WORKFLOWS?

Workflows are the processes that power your business. They can be as simple as writing an e-mail and as complex as building a car.

Here's how I explain it:

A workflow is a sequence of non-overlapping steps that define how a piece of work moves from start to finish.

Workflows are typically described as the sum of their parts:
- Resources put into the workflow, including: people, time, money, parts, etcetera
- The steps taken from beginning to end, including prep work
- Outcomes of the workflow, usually described as the end product

In this book, we're most concerned with significant workflows. These tend to be more complex processes, or those that occur frequently, and have major implications for the business. (So, usually not the once-a-month processes that take fifteen minutes). They offer your organization the most return on the investment of revising and mobilizing them.

Examples of these kinds of workflows include:

- Two workers loading grain into a truck
- A sales rep converting a lead into a customer
- A field service person installing a cable modem

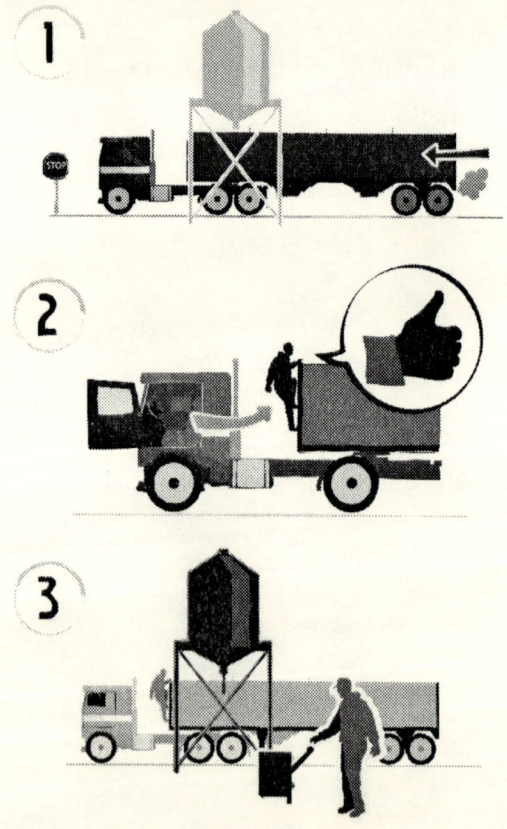

All elements of a process make up a workflow. Your business has many of its own workflows. Each and every workflow that runs through your business could be better. It doesn't matter if you're the most successful company in your industry—or even on the planet. Processes can always be more efficient. Make sure you focus on the ones with the highest potential.

Let's start with the simpler ones—workflows that are easy to identify as inefficient or flawed. A solution exists. A primer in mobility is what you need to find it. How about some practical information you can use right now?

Used to redefine workflows, Billion Dollar Apps are:

Clipboard Killers

Many workflows are still paper-based across industries. These systems *must* go mobile. Billions of dollars are at stake.

Paper workflows have two main flaws. First, they're unguided, meaning that workers must rely on game-time logic (they have to know *exactly* what they are doing). For example: *If someone orders product A, I must make sure to ask about product B.* This idea has to be trained and reinforced. It's tougher and tougher to rely on people as workflows grow more complex. With mobility, guided workflows ensure workers wrestle with the task at hand, not the process.

The second flaw is double data entry. The worker writes information down, only to have to transfer it to the electronic system later. This allows the introduction of data errors and other inaccuracies, not to mention hours upon hours of wasted time. Mobility means you enter data once, reducing errors and saving time. Anywhere you see a clipboard, binder, or notepad in your business, you see an opportunity.

Hardware Enablers

Warehouses, plants, grain silos—wherever complex hardware resides—beg for app integration. Workers can control every one of these systems using Wi-Fi and the right app.

In some cases, businesses could even replace that hardware. Urban Outfitters is a prime example. They asked the question, "Why should shoppers have to come to the cash register to buy?" The result was an iPhone case that works with the phone (via an app) to scan bar codes and process credit card transactions. Associates handle customer transactions anywhere in the store.

Access Shortcuts

Data has always been important. Now it's playing an even bigger role in business. Unfortunately, recording and accessing data are extraneous steps in workflows, standing in the way of true efficiency.

The BDA approach is more than just a clipboard killer. It's a step saver. It brings business closer to real time, reducing the time it takes to accomplish the goal.

What if your field sales team could take someone from lead to customer during the first meeting instead of the four meetings it usually requires? What if you could enter data as soon as it's received and generate a customized solution?

This is not the stuff of the future. This is now, and it's available to your business if you're willing to make the investment.

Kiosk Replacers

The self-service revolution has reached its next phase. Anywhere your business uses kiosks, it's time to move on. It's time to bring mobility into the picture. Kiosks are stationary, expensive, and limited. They keep customers in one place instead of wandering your aisles. They drag employees away from tasks to get information they could just as easily access from a mobile device. Any of this sound familiar? These touch on pain points that plague most organizations that haven't adopted mobility measures.

A MORE EFFICIENT ENTERPRISE? THERE'S AN APP FOR THAT

What about all that technology you invested in over the last decade? Are you supposed to throw it away?

Mobility is also about breathing new life into old systems. No one wants to use a customer relationship manager (CRM) that's complex, confusing, and clunky to manipulate. An app can make it engaging, understandable, and far easier to use.

The business world barely bats an eye when a Fortune 500 organization like Sonic Automotive invests $57 million in technology in a single year. A large portion of that cash went to mobility-based technology. Sonic executives want to revolutionize the company's sales process.

For journalists, the Sonic story is led by Apple, iPads®, and big bucks. Read further to find the underlying narrative: how Sonic redefined its workflows and gave new legs to old technology. The company built a suite of apps that acts as a window into their dealership management system (DMS), redefining how personnel use it.

For Sonic's sales associates, it doesn't take long to figure out that the real value is in how an easier DMS process leads to more sales and higher commission payouts.

Executives are happy, too. Sales reps are doing more business. They didn't have to shell out an arm and a leg to replace the old DMS, either.

So why don't our systems work well enough in the first place?

We jumped too far too quickly. We gobbled up technology before it was ready for the people who need to use it. Now many organizations are paying the price. Some don't even realize it. It isn't the systems themselves that cause problems; the interfaces just suck. Your technology platform could be an industry game-changer behind the scenes—but if it isn't user-friendly, it won't return on its promise. Users just don't have the patience to fight through complex systems. This means that they can't take full advantage of their capabilities.

BILLION DOLLAR APPS ARE THE ANSWER

BDA is a workflow revision process that helps you find the pain points with those old systems and redefine how end users interact with them. It's a conduit for simpler, smarter interactions.

Mandating a complex and confusing platform is never the answer. You'll either waste employee time or have trouble getting buy-in from workers.

Creating a new window to that platform, on the other hand, reboots the presentation while maintaining the working parts. It makes inputting, accessing, and analyzing information easy again by improving the user experience (UX).

A strong mobile app is like a window to your legacy system. Refitting that system for mobile offers benefits like:

- Quicker processes and digital workflows through making them more intuitive
- Recouped money and time lost from data-entry mistakes
- More comprehensive data capture, making workflows stronger
- Workers who are empowered to access info from anywhere
- Happier, more efficient, and more effective employees

You have powerful platforms at your disposal. Making them more appealing for the worker can have a huge impact on your organization. Revamping legacy systems is an important benefit of mobility. But could it apply elsewhere, too?

WHAT ABOUT NON-TECHNICAL WORKFLOWS?

Billion Dollar Apps aren't limited to legacy systems. In fact, infusing non-technical processes with technology is one of mobility's biggest game changers.

Neil Mylet did just that. The grain silo had no associated web system, no laptops in truck cabs, no backend enterprise data platform. It was a complete no-tech environment—exactly the kind of ecosystem where mobility will have the greatest impact.

Apps help businesses everywhere take back control of workflows. Anywhere workflows haven't been infused with technology is an opportunity.

It's tough to ignore the benefits of mobility, regardless of your business's current technical situation. The stakes are high. How can individual workers and entire businesses benefit from mobility?

WHAT YOU STAND TO GAIN

Billion Dollar Apps is an intentionally provocative title. It's also *real*. An industrial services company saves millions of dollars a year simply by streamlining invoicing. A health insurance firm closes 300 percent more business by simplifying how customers submit claims. An energy company shaves two thirds off the time it takes to conduct the auditing process, a workflow completed many times weekly by field agents. All of that because of Billion Dollar Apps at work.

Saving time and money while increasing revenue potential is what enterprise mobility is about. But these are the overarching benefits. To understand how you get these, we have to dig deeper and find the granular business benefits.

HOW LOWE'S TOOK THE INITIATIVE WITH CUSTOMER SERVICE

Two years ago, walking into a Lowe's for the first time was overwhelming (especially if you were just there to pick up a box of nails). Like other home megastores, it's not quite your neighborhood hardware shop.

Now, thanks to mobility, the Lowe's customer experience is a little less overwhelming—and a *lot* more effective. The company recently mobilized its workforce, rolling out three apps and forty-two thousand iPhones® to curb major workflow inefficiencies.

Lowe's execs revealed major challenges in their customer service workflow. Deconstructing processes helped them identify challenges like:

Customers weren't getting what they needed quickly enough.

Before the app, Lowe's reps would engage customers to help them find products. Personnel had to visit a kiosk to find inventory and availability. Long wait times often meant waning customer interest. If items were out of stock, shoppers left, likely visiting a competitor. Lines at registers meant an even longer time investment for customers.

Managers spent countless hours on administrative work.

Spending hours in a back office pulling and printing reports is no one's idea of a good time. That's just what Lowe's managers did daily. That time would be better spent on the floor.

The backend system was ugly and hard to maneuver.

In-store workers keyed SKU numbers and UPC codes into a nondescript database. The system returned inventory in a format that was tough to read. The system was not intuitive and caused even longer delays when floor personnel served waiting customers.

Addressing these problems led to the creation of three apps: two for store workers and one for the customer. Each served to fill these business needs.

Employee app

Each customer service rep's iPhone is equipped with an inventory scanner and credit card processor (just like Apple stores). Besides processing transactions, the app pulls product information, how-to videos, and more at a moment's notice.

Manager app

Inventory, price audits, sales reporting, and more are just a click away in the manager's app. The tool cuts out the back-office hours and saves on printing resources.

Customer-facing app

Like the employee app, the customer-facing app empowers shoppers with information about the Lowe's location and inventory the store has in stock.

Planning out the backend had to come first, before the organization could even start planning the customer-facing app. Strong app design made that ugly backend a much smoother and friendlier user experience.

Like Lowe's, your organization can make giant leaps forward by redefining business processes and enterprise mobility.

THE BENEFITS OF REDEFINED MOBILE WORKFLOWS

Picture your company with happier employees. Imagine a thriving, bustling organization—all thanks to a new approach to how you do business.

These sorts of results make heroes in organizations. I've been privileged to help a slew of leaders become the change agents their teams needed.

You'll need to build a business case. Every good business case leads with the benefits.

The bottom line

First and foremost, management wants to know what kind of impact mobility has on the bottom line. You already know that apps mean a huge savings on time and money and that more efficiency leads to more revenue. What you'll need to know is how to calculate it. (We'll learn about that later on.)

Happy, productive workers

The right apps make the job of the individual worker that much easier. Apps make the biggest impact when they zero in on a pain point experienced by the workers actually completing the task. This perspective is critical to a successful mobility project.

Making a worker more productive means more than just improving the bottom line. It creates happier, more efficient workers who are ready and willing to do more.

More efficient business

Everyone wants a more efficient business. Mobility brings that kind of efficiency, empowering workers with anywhere access to critical technology.

Competitive advantage

More revenue, fewer costs, happier workers, and more efficient business add up to a major competitive advantage in your marketplace. We're past the very early adoption phase of business mobility, which means you reap the rewards of mobility without the pitfalls and false starts.

Better platform adoption

No more half-assed attempts at getting your employees to adopt a new platform. People should be compelled to use your technology, especially in the workplace.

Better platform adoption means workers won't take shortcuts. It means more comprehensive usage of your platforms. Your technology works best when you use it the way it was meant to be used. With mobility, you get the most out of your technology.

Best-laid plans

Got vision? Smart business targets the future. Today's technology moves so fast that one misstep becomes a huge drag on even the most successful companies. Adopting mobility now ensures you'll have solid footing for years to come. We have a good start. But you'll need more details to build a truly powerful business case.

Ready to become a tech pioneer? Mobility isn't exactly a well-worn path. Like any good explorer, you'll need to make sure you don't lose the road. After all, there's a valley between the *obvious* app and the *right* app…

OBVIOUS APPS VERSUS RIGHT APPS

The obvious app isn't always the right app. In fact, it rarely is. When mobility really took hold, all of the fuss about apps fell on the consumer side of business. Therefore, businesses considering mobility often go straight to the consumer.

One of the most obvious customer-facing apps you can create is one that sells. Tons of companies fall into thinking that since they sell stuff through their websites, they should use their apps for the same purpose. An app focused on marketing or product sales is undoubtedly the *obvious* app. But does it deliver any real value? Is it truly the right app?

Every business should ask itself these questions:

- Are we just moving customers from one purchase channel (the web) to another (mobile)? Or are we adding new value?
- Why are we building this app?
- Does it solve a business need?
- Is that need a pressing one?
- Does addressing it offer the biggest bottom-line impact?
- Are we looking at every component of the problem?

Obvious apps aren't always implemented on the customer side, either. It's a common pitfall for an organization to take existing web systems and move them to a smaller screen. This kind of app doesn't take into account the different features and workflows mobile workers need to get the job done. It's very likely that someone using an app in a mobile setting needs to do different things in a different order than the person sitting at a desk.

Build a relationship with your audience through mobile capabilities and they'll reward you. Take metal and supply chain solution distributor A.M. Castle, for example. The organization's inside and outside sales teams function differently from each other. The obvious apps would have focused on the people in the field:

the outside sales team. A price quote tool, for instance, seemed like a great resource.

But a Billion Dollar Apps evaluation showed that these apps meant very little in terms of making work easier for the outside sales team. What's more, they would have had little effect on the bottom line. (These two outcomes are usually related.)

Instead, A.M. Castle decision makers did the work on proving the value of their apps. They found that the biggest impact could be made equipping the inside sales team with better tools. A mobile-first redesign (and subsequent suite of apps) for the organization's slow inventory and order management web app was the better option. It cut time spent using the web app by a startling 96 percent.

The obvious app doesn't impact business the way the *right* app does. The obvious app isn't the Billion Dollar App. To find the app that's right for your organization, you'll need some guidance.

BEFORE YOU GO MOBILE, FINISH THIS BOOK

We can't all be developers and mobile experts. But that's not who is reinventing industries right now. You can be an agent for change. Spearheading your own Billion Dollar Apps can be a career-maker. Change agents like you have a uniquely double-sided perspective. You understand the current technology infrastructure. You've probably even experienced some of the old systems. But you also stand on the horizon line, where future technology has begun to come into focus.

Few executives have the benefit of that perspective. It's your role to make their jobs a little easier and build the business case.

You'll encounter challenges, including:

- Onboarding your executive or other decision-making team
- Choosing the right apps for the job
- Deciding how far to go with your project
- Determining the order in which you should build your apps
- Picking a development partner
- Figuring out how to measure success

Throughout this book we'll address each of these challenges. We'll see exactly how they manifest themselves throughout the Billion Dollar App process. We'll see how organizations just like yours have successfully overcome them.

Fostering mobility in your organization takes time and dedication. A guiding hand is crucial to the process.

WHY SHOULD YOU LISTEN TO *ME*?

I've been applying technology to solve business problems since graduating from the University of Illinois in 1993. Since then, I've founded eleven businesses. Each one has been dedicated to applying technology to help people thrive, especially in a business setting. Enterprise mobility is something my company, Lextech, does daily.

In the past, I've used awful apps with no real purpose. They're frustrating, especially if they're acting as roadblocks to getting the job done instead of conduits. The information in this book should be in everyone's hands so that we can collectively rid the world of purposeless apps and start driving serious value.

I've seen workflow innovation make an amazing impact on companies big and small. The team at Lextech has made it happen. Building Billion Dollar Apps is our business.

Many of the anecdotes you'll find in this book are based on firsthand experiences, even the negative ones. I'll use real names wherever possible. In others, confidentiality agreements and business tact limit me from identifying those involved.

Know one thing: These are true stories of the successes and failures of mobility's early adopters. With this collective experience, we'll walk the path to process improvement through mobile business, step by step, until you can easily identify your Billion Dollar Apps.

To get there, let's first get up to speed on where we've been and where we're going.

CHAPTER 2
How Mobile Changed the World

Mobility has completely changed how we interact with information, applications, and each other. Most of us remember a world where sitting down at a desk was necessary to get the information we needed. More recently, we could pull that information up on a laptop wherever we were—if we were willing to lug it around.

Throughout the time of desktops and laptops, we had cell phones—devices we could carry around to connect with others on demand—but we lacked the ability to connect directly to information.

Are you surprised that we often called people who had access to the information we craved? "Phone a friend" wasn't just a call to action for aspiring television millionaires. It was a daily occurrence that helped lead us directly to an era of mobile Internet access.

YOUR FIRST PERSONAL COMPUTER

That hunk of metal and microchips in your pocket isn't a phone. It's a personal computer—the world's *first* personal computer, actually. Coined way back in the '70s, the phrase "personal computer" is more of a marketer's slogan than anything else. A truly *personal* computer is with you at all times. Your smartphone is the only personal computer you've ever owned.

It's also an extension of your personal story, a window to your digital life, and a professional productivity tool. Today's smartphone owner can hardly remember a time when she couldn't watch last night's *Walking Dead* on the train ride to work. Come to think of it, neither can I.

It's hard to imagine a world without mobility driving our daily activities. Being out of touch for an extended period of time is now a very real phobia. If I forget my wallet at home, I can probably survive the day. If I forget my phone, I'm turning around to go get it.

That's the kind of impact mobility has had on our world: we just can't live without it. This shift happened in less than a decade. As we'll see, the rise of mobility comes from the empowerment of consumers. Consumer demand has shaped the development of our mobile world, and business mobility is at its behest.

What's going on? Why are so many people flocking to mobile? Why hasn't business led the charge?

All of this is because the iPhone (and every device it spawned afterward) is so easy to use that a baby (or, as in my case, a tech-newbie mother-in-law) could do it.

HOW EASY IS THIS?

My mother-in-law never used a computer in a work setting and didn't have much interest in the computer we set up for her at home. But after a few minutes with an iPhone, she was hooked. The first thing she did was download an army of apps. She interacted through social media for the first time. The iPhone introduced her to the world of casual gaming, and of course, it became her camera of choice for snapping pictures of her grandson.

Getting an iPhone also drove her to put in a home Internet connection with faster access and a Wi-Fi network. Suddenly steeped in technology, she loved the new iPad my wife and I got her for a gift. The larger screen made it even easier to interact with the apps she loves.

Why the shift from disinterested onlooker to passionate user? Her computer was intimidating. From using a mouse to picking from dozens of options while word processing, there were so many bits and pieces to figure out. Using the web wasn't much easier for her. Most users—*especially* my mother-in-law—only want to perform a few specific functions on a website, but developers are compelled to cover every possibility in one place, making it overwhelming for new users.

The iPhone (and mobile in general) presented two new, friendly concepts: *touch* and *apps*. The touch interface is awesomely easy to use: so easy, a baby can do it.

Touch interfaces are one of the most important developments of the twenty-first century thus far. They don't require any sort of manual. All you need to start interacting are your fingertips—instruments you've been using since the womb.

Web apps just aren't designed for touch interfaces. Their user experiences need to be rethought.

Well developed mobile apps are sharply focused on one or two key tasks, making workflows intuitive and simple, even for the most tech-challenged users. They don't require any training or user manuals. Someone who has never used a computer in a corporate setting is totally comfortable with them.

Mobility offers the flattest learning curve in the history of technology. Apple saw the opportunity to strike while the iron was hot. Before mobility really started to sink into the business world, the iPhone debuted as a consumer tool first. It was successful because it was the most intuitive smartphone experience on the market.

You could make the argument that the Blackberry has been a core business tool for a long time, but it never got beyond the basic communications of e-mail. It was Apple that pioneered the use of apps to turn the phone into a smartphone. The company set the tone for business mobility.

Adopting enterprise mobility now is no longer just about getting ahead. You'll want ample time to prepare a plan. If you don't start soon, you probably won't keep pace with the competition.

MOBILE ADOPTION HAS REACHED ITS TIPPING POINT

How many people do you know who *don't* carry mobile devices? You may still know a few stubbornly holding on to their old cell phones. But the majority of your friends, relatives, and coworkers carry smartphones. I can safely make this assumption because I have data to support it. Mobile adoption has reached its tipping point, making mobility the new norm.

In February 2013, comScore revealed that smartphones had surpassed 50 percent market penetration. The organization calculated that 125 million Americans now own smartphones, and 50 million have tablets.[1]

This news meant that we've reached the "late-majority" stage of the technology adoption curve. We've also passed the mobile "tipping point," the moment when mobile device access becomes typical among Americans. Mobility's momentum will carry it into an era where it will be the norm, a dominant force in how we access information. Here are two more eye-opening stats about the future of mobility.

Desktop usage for web access dropped 5 percent over the six months leading into 2013.[2]

---

[1] http://www.comscore.com/Insights/Press_Releases/2013/2/comScore_Releases_the_2013_Mobile_Future_in_Focus_Report

The drop was widely attributed to the combined 12.5 percent gain in mobile device usage. This means mobility is displacing desktop usage. It is, in fact, forcing the desktop toward extinction.

By 2015, the global mobile app market is expected to reach $25 billion.[3]

That's no slouch of a number—and it doesn't even take into account the private developments happening across global enterprises. Mobile engagement services as a whole are expected to reach $32.4 billion by 2018.[4]

> Laptops and desktops will stick around for power users who create deep content or write software, but for 95% of users they won't be necessary.

These are powerful statistics that support the assertion that mobility will one day replace most desktop computers. That day isn't as far off as many business leaders think. Rapid consumer adoption is telling us something, and it's shaping the face of the mobile industry.

---

[2] http://www.emarketer.com/Article/How-Do-Internet-Users-Divvy-Up-Their-Desktop-Mobile-Web-Time/1009841
[3] http://techcrunch.com/2011/01/18/report-mobile-app-market-will-be-worth-25-billion-by-2015-apples-share-20/
[4] http://www.forbes.com/sites/forrester/2013/08/09/mobile-engagement-providers-will-be-a-new-32-4-billion-market-by-2018/

WHAT EFFECTS HAS SWIFT CONSUMER ADOPTION HAD ON BUSINESS?

Consumer trends are the most powerful driving force behind today's mobility. Consumers place demands on businesses that define how they want to interact with their companies.

In some cases, this results in business models that cut out "middlemen" and create a direct link to the consumer. But it's mostly about consumers being in the driver's seat. That mentality carries over into the business world, where employees are conditioned to expect easy mobile access to enterprise systems the same way they access consumer tools.

End users (our customers and employees) continue to get the final word on mobile development. Here are four ways they've driven progress.

End users shape purpose.

There's no doubt the general public has had a lot more time to work with mobility than the business world. As a result, consumer opinion has had a major impact on how the world of mobile productivity has advanced over the last few years.

This isn't necessarily a bad thing. Sometimes the business world needs a nudge to sacrifice formality for usability. Major enterprises have already begun to take the reins, kicking off a rich era of mobile productivity.

Business associates mobility with distraction.

Some execs still see a need to keep mobile devices "off the network." There are a few regulated industries that have to deal with this issue, but primarily, this perspective is wildly outdated.

Associating mobile with unproductive use has held many organizations back from exploring the medium's true benefits. In the same vein, treating it as an "always-on" communication tool is missing the point. It's a tool that employees can and should use *while* on the job—not just for getting a quick answer on a Saturday afternoon.

Apps face consumers.

Mobility has also had a tough time extending beyond the walls of marketing. That's the go-to these days: "Consumers like apps. Let's give them one that makes accessing our products easier."

Selling directly through an app is usually the last thing you want to do. Consumer-facing apps must link the consumer into a workflow that builds his or her relationship with the organization. Connecting consumers with experts who can provide valuable answers to important questions, for instance, is much more valuable than giving them an interactive catalogue and expecting them to buy. Just moving someone from the web to a mobile interface isn't adding value unless the workflow and use scenarios are really evaluated and understood.

Consumer-facing apps must be focused on building long-term relationships. This requires brands to push apps with tools and content that consumers can use regularly instead of just highlighting product-driven "look-at-me" messages. Most companies could use that shift in perspective.

Millennials expect more.

A new workforce generation demands mobile tools as the norm, and businesses must answer the call or disappear into obscurity.

The fact is, the younger generation is growing up with tablets and smartphones, not desktop and laptop computers. Millennials aren't sure what to do with Internet Explorer, much less a floppy disk. Organizations simply won't be able to hire or sell to Millennials without mobile devices and apps. In today's business environment, that's a risk you can't take.

Consumer behavior has clearly had a major impact on mobile trends. The future of mobility relies heavily on the sculpting hand of the consumer.

BUSINESS MOBILITY PAVES A NEW ROAD

Experts predict tablets shipped for use in the enterprise setting globally will reach 96.3 million units by 2016. Add smartphones to the equation and that number goes through the roof.

The question you must ask of your organization is whether or not it plans to be counted among the ranks. It's that one giant leap that will drive your company forward. And it can come from anywhere.

Throughout this book, you'll see lots of examples of how organizations are setting their workforce up for success through mobility. But what we'll also see is that tailoring mobile tools for your customers can factor heavily into how mobility affects workflows.

The expectation that your app must face the customer can be used to your advantage if the resulting efficiency benefits an enterprise workflow. This is exactly how we collaborated with PayFlex, for example. The result was a better customer experience that also boosted a major revenue-generating enterprise workflow. Working together, these make a powerful argument for mobility.

EASING THE PAIN OF PROCESSING CLAIMS

PayFlex is a company that gets it. Their quick success, according to an exec, is almost entirely rooted in mobility—a true case study in how Billion Dollar Apps can make or break a company. In 2011, the insurance giant Aetna recognized its success and purchased the smaller health-care finance business while retaining its brand.

A little background: PayFlex is rooted in health savings accounts (HSAs) and other account-based health plans. The organization partners with companies to supplement benefits packages. The end user is the partner company's employee.

It all started with an internal process: how PayFlex processed claims. This specific workflow required cooperation with some unlikely workers—somewhere around one million end users. PayFlex could only process claims as quickly as customers sent them. This process relied on subscribers to mail a physical receipt after leaving the service provider's office.

As you might expect, most people weren't receptive to that process. To make the workflow more efficient, PayFlex needed buy-in from their customers. HealthHub, the organization's existing web system, offered the ability to upload receipts. Still, it didn't have the convenience factor consumers wanted. It had to be even easier.

The idea for the HealthHub mobile app was born. Customers snap a picture of the receipt, sending it directly to PayFlex's team for processing.

According to PayFlex, the app boosted their sales close rate from 32 percent to an astounding 96 percent with enterprise customers. It also generated media attention, eventually leading to its acquisition by Aetna. PayFlex is on the cutting edge in an industry that's been slow to react to workflow technology. Investing the resources to build its first app showed a huge return.

Quick adoption of the company's HealthHub app by consumers exemplifies a stark reality—mobility isn't just a luxury. It's an expectation.

PayFlex understood that app development is *not* about putting a mobile face on your website or your catalogue. Mobility is much, much more than part of the cost of doing business. It's an opportunity to cut costs and earn revenue. It's process improvement and workflow reconstruction. It required adding capabilities that made sense in a mobile setting—receipt capture at the point of action in the pharmacy.

Billion Dollar Apps isn't just a catchy book title. This growing trend of mobility is an opportunity to think bigger about your business.

GO BEYOND THE OBVIOUS

An electronics manufacturer recently approached my company for a mobility project. They had a very clear plan—to put mobile interfaces on their products.

That was an obvious step. We tried to get the company to ask the bigger questions: *How can we apply mobility to advance our business? How can we approach our customers and marketplace differently than just selling widgets? Can we couple our expertise with our products to offer our customers even more value?*

We tried to help them look beyond the idea of mobility as a nice thing to have for their customers. To truly leverage mobility, they needed to think about what it would take to develop apps that *create revenue* and support their other company goals. They considered mobile part of the cost of doing business and wouldn't look beyond that initial mobile app to talk to their equipment, neglecting to answer *why*.

Your mobility project is destined to add little following this approach. Without a truly goal-oriented plan, your project will fail. What this company wanted is what I mean by *obvious apps*. Thinking bigger means looking beyond consumer mobility. It means looking inside your organization and asking how you can do things better. In essence, the success of enterprise mobility relies on choosing the apps that are *right* for your organization. It means enhancing those workflows with tools that make sense in a mobile setting and add value.

The electronics company's resellers were service providers that could have benefited from mobile tools designed to make site inspections easier. Following the inspection, whose product do you think the reseller would suggest for the customer?

What about a more compelling way to engage customers? The reseller's field agents could have used video and other multimedia to convey the importance of certain products and services to their customers. A tool like that might have helped agents upsell, translating into more sales for the electronics company.

All in all, they had a great opportunity to provide a simpler, more valuable experience for their resellers—an area that could have had a much bigger impact on the business.

Identifying the right apps requires a proven process. Next, we'll see how this process worked for Protectus, a hypothetical insurance firm.

CHAPTER 3
How Protectus Discovered Billions of Potential in One Year

The right apps can transform a successful business into an incredible one. That's exactly how Protectus became a powerhouse in the world of insurance.

Is it possible for an organization to nearly *double* revenue in a single year? Can you *really* cut time spent on workflows to fractions just by adding bridges between the customer, your sales reps, and your data?

Protectus did. So have hundreds of enterprise organizations across the spectrum of industry. Each found success because they focused on the right apps for the right reasons. The Billion Dollar Apps process nestled between the pages of this book will help you get there too.

Disclaimer: Protectus isn't a real organization. But its story is based on dozens of companies I have seen make the same transformation. We'll use Protectus as a template to illustrate each part of the app strategy and planning process.

Let's take it from the top: What does Protectus do?

ABOUT PROTECTUS

Protectus, Inc. sells insurance and financial products for individuals and families. Over twenty years, the company has grown to $5 billion in annual profits. It has exhibited an average of 3 percent annual growth since its inception and competes well against Internet-based insurance providers like GEICO and Esurance.

Around a thousand sales people work under the Protectus flag. These sales reps are tasked with driving organizational growth through business development.

The company has identified ambitious revenue goals for the new year: a 10 percent growth in top-line sales. Achieving this goal will establish Protectus as a leader in its industry, but doing so will require much more than a bigger or more motivated sales staff.

Protectus needs a hero—someone within the organization who understands how the business works, stays highly motivated, and is capable of identifying and implementing change. Someone like Samantha Lee.

ABOUT SAMANTHA

Samantha Lee joined Protectus five years ago as a seasoned sales rep. During her first two years, her strong performance in the field led to a promotion to regional sales manager.

Protectus management saw her potential early in her time with the company. The vice president of sales, a woman who sees much of her younger self in Samantha, took a particular interest in the promising saleswoman. They formed a close relationship. Sam sees the vice president as a mentor and friend.

Sam's hard work paid off. She was named director of field sales at the beginning of the year. Now she heads a team of a dozen regional managers. Sam is a **change agent**—an ambitious person in a position of influence who understands the need to adopt innovative approaches within her industry.

Put simply, change agents effect positive change in their organizations. Change agents tend to have strong track records for innovation. Sam is no exception. With Protectus, she has successfully implemented a new sales training program, among other improvements.

People just like Sam (perhaps just like *you*) drive innovation at enterprise businesses across the globe every day. Often they meet with seemingly insurmountable challenges, including limited resources, change-averse managers, and priority conflicts.

Sam is the ideal change agent: experienced in the field, concerned about the future of the organization, and determined to effect the kind of change that's worthy of its own line on her résumé.

For nearly five years, Sam was on the front lines of business development. She gets the needs of Protectus customers. She understands the challenges of her new team and empathizes with them. Creative thinking got her this far. With Sam's help, Protectus will usher in a new age of growth.

THE FIRST MEETING

Early in January, the vice president of sales calls the entire sales team into a meeting. Team members located across the country join remotely. The meeting is for going over the new year's goals put forth by management. "This year, Protectus' goal is to increase top-line revenue by 10 percent," the VP explains. "The burden of that growth will fall to us in sales."

Some of the more seasoned managers in the room seem dubious. They watch Sam, a longtime peer, for a reaction. She stays calm and nods her head with the presentation. As director of field sales, she's been briefed on the new goals. But she doesn't have a solution just yet, either. How can the sales team boost numbers without adding more people? They're already working long hours. They need help.

Unfortunately, that help doesn't look like it will come from the executive team— not without a strong business case, at least. Sam has met her CEO on a few occasions. He is passionate about his business. He has a strong vision for the future and needs his team to figure out how to execute to get there.

But Sam also understands the limitations of vision that come with such a lofty position at a large company. The CEO can only see so far into daily operations. He can't accompany sales reps on their visits with prospects. An accurate view of the customer relationship management (CRM) software's limitations is only possible when you work with the program regularly.

Sam sees her team's task as a challenge. Of course, she's never backed down from one before. She resolves to find a solution and assures her sales team of her conviction.

Sam's industry media feeds have been packed lately with technology-driven innovation. Could technology be the answer? She's wondered about this before. The sales team's CRM is its primary technical platform. Could there be a way to make it better, more effective? Right now, it doesn't pack the punch the vendor has promised.

A CRM is a system that helps salespeople track customers and potential customers. The platform logs sales information like customer correspondence, order entry, new opportunities, and sales reporting. Sam remembers its debut just four years ago. The IT department made a major push for it, working with sales managers to customize the solution and offering in-depth training to team members.

The man behind the CRM rollout is now the director of IT. If anyone knows the strengths and weaknesses of the software, it's him. Together, they could brainstorm improvements to the software configuration. Later that day, Sam sets up the meeting.

MEETING WITH THE DIRECTOR OF IT

Although Sam is up to speed on the latest consumer technology, she doesn't consider herself very technical. Fred, the IT director, could be a valuable resource and ally.

She explains the situation over coffee. "I feel like we aren't getting the most out of the software," she says. "People are taking shortcuts, leaving out information, making errors, entering duplicate information. Maybe we could make the platform a little friendlier for the team. What do you think?"

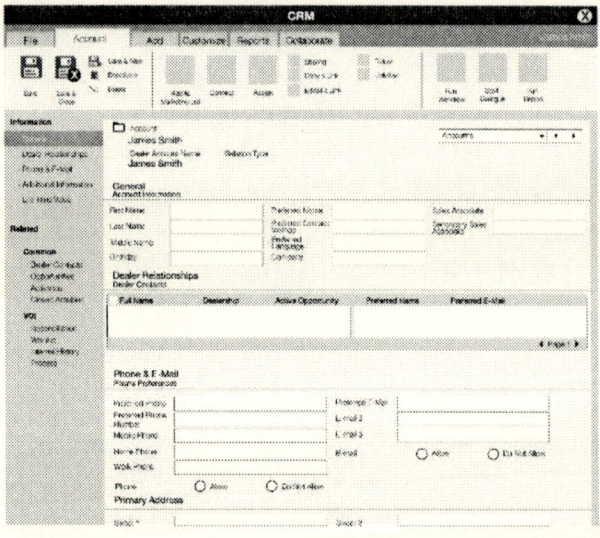

A man of pride, Fred doesn't see anything wrong with the software. "I'll tell you what," he offers. "How about we do some more training?"

Sam smiles. "I don't think it's that simple, unfortunately," she explains. "We train every new rep that comes through the door, and they're all having the same problems. Besides, that doesn't really make it easier to use outside of the office. And it won't help curb the data-entry mistakes that reps are making."

"If they entered data immediately after meetings, you'd have fewer issues," Fred counters. "The information would be fresher. That's why you guys have laptops. That's why the system is web-based."

Sam explains how opening a laptop in the middle of a meeting can intimidate the customer. "We don't always have a great Internet connection when we meet with prospects, either." She takes the opportunity to broach the subject of mobility. Fred listens intently as Sam describes her experience with the CRM vendor's mobile app.

Unfortunately, she explains, the app isn't very easy to use and doesn't seem to fit the unique workflow needs of her team. It's almost as though the app was an afterthought for the vendor.

Fred acknowledges that mobile access is useful and that the IT team agreed that the mobile app wasn't useful. That's why they've already spent time making the CRM's web portal mobile-friendly.

But Sam argues that it isn't enough—that the system needs to be more accessible, more compelling, and more intuitive.

"That'd make accessing it more congruent with the current laptop," Fred admits. "How about we use a remote desktop app to connect to the rep's main computer?"

"I just don't think using a touch interface to drive our laptop back in the office works well. The tablet and laptop screen sizes are really different, and they seem like they're meant to be used differently," Sam explains. "It can get clunky. Plus, it still relies on having an Internet connection."

Fred steers the conversation away from long-term change toward quick-fix ideas that won't require process revision. It's clear to Sam that she'll have to go back to the drawing board. In fact, she may have to work things out without the help of her IT director.

**

We can blame Fred for closed-mindedness and maybe a bit of bias. He suffers from departmental blindness that's typical in organizations that work in silos. He lacks visibility into the sales process. Tough problems appear to have simple solutions.

Fred's mind-set is rigid: he isn't thinking about process change. He doesn't understand the power of mobility. Like many other managers at his level, he sees tech adoption as a requirement of the job. Why should the organization bend over backward for people who aren't performing their responsibilities?

Not to mention, he was the architect of the current system. Fred may not be an unreasonable person. But he is protective of the platform he helped assemble. After all, he's always led tech innovation internally. What does Sam know about it?

For now, she will have to find a solution without his help. She may not consider herself technically minded, but she does understand that technology plays a critical role.

Perhaps just as important is her ability to dream big. After a few days of brainstorming, the bud of an idea starts to form. Over lunch, a friend (and ex-coworker) raved about a new tablet-based sales system that made her job so much easier.

Sam loves her iPad. She always hears about how tablets are making waves in the business world, but she isn't sure how. Could her friend have the answer?

A BILLION DOLLAR APP DEVELOPMENT PROCESS

Sam tells her story to her friend Angela, a regional sales manager at a noncompeting company, who nods understandingly. Angela was directly involved in the project Sam is most interested in, so she has a strong understanding of what needs to be accomplished.

Angela explains that Sam's situation is almost identical to the one that spawned change at *her* organization. Sam is right: the sales team needs tablets. But the enabler, the key to unlocking the tablet's potential, are custom, proprietary apps that tie back to Protectus' CRM. "And you'll need the process we used to make sure you do it right the first time," her friend adds.

Angela explains the six-step process for finding Protectus' Billion Dollar Apps and promises to forward along the workbook they used to get there. The Billion Dollar App strategy focuses on a single end goal: create a prioritized list of the right apps to use as the road map to mobile success.

The steps are:

- Take a good, hard look at your business strategy.
- Analyze the workflows that support your business strategy.
- Identify the issues that make these workflows less efficient.
- Develop concepts for apps that will solve these issues.
- Calculate the return on app, including new revenue and cost savings.
- Score the app concepts against each other to figure out which are the most valuable.

Angela shares detailed resources to help her out along the journey. The next day, Sam begins putting them to work.

STEP 1: UNDERSTANDING THE BUSINESS STRATEGY THAT REQUIRES CHANGE TO SUCCEED

To get started, Sam sits down with the year's goals handed down by the executive team. Here's what she knows:

The current top-level strategy for Protectus is to **grow top-line revenue by 10 percent this fiscal year**. The strategy behind this goal is to **close more sales**. These goals must be accomplished through the current sales force.

Sam identifies the following key drivers for accomplishing this goal:

- Boost the number of presentations without adding personnel.

- Increase the close rate.

The sales department is mostly responsible for accomplishing this goal. Sam's next step is to examine the workflows that define Protectus' sales process.

STEP 2: ANALYZING SALES PROCESS WORKFLOWS

Becoming a top sales rep at Protectus requires a strong understanding of the company's portfolio, excellent communication skills, relentless days of meeting offsite with potential customers, and late nights spent at the computer.

Here's how the sales process typically plays out.

48 | BILLION DOLLAR APPS

Day 1: Starting from Scratch

During the first days of the cycle, the sales rep spends time getting in touch with potential new customers. This part of the process may consist of calling and emailing cold, contacting existing customers for a catch-up, and following up with warm leads.

The goal of this first portion of the sales process is to schedule meetings. The salesperson books an initial meeting with a prospect for two weeks later and prepares her collateral.

Day 15: The Initial Meeting

The sales process continues in a face-to-face meeting with a potential customer. The rep hopes to gather enough information to create a viable proposal.

Lasting one to two hours, the initial meeting is a chance to spend some time learning about the new person—details like personal background, financial situation, and dreams for the future. The prospect fills out forms. The sales rep asks questions and takes notes.

Some sales reps write down information in a notepad. This particular rep opens her laptop to take notes. After she learns about the prospect, she shows him or her a short PowerPoint presentation introducing Protectus.

The rep thanks the prospect for the meeting and schedules a follow-up meeting if the prospect is interested. She heads back to the office and spends time manually transferring the forms and her notes into the CRM system.

Now she has to create a personalized proposal for the next meeting, choosing the products that make sense for the prospect based on the personal information she's gathered. She spends time in front of the computer, building scenarios based on information like marriage and family status and the age at which the prospect wants to retire. She consults with Protectus product experts to find specialized solutions.

Born from all of this research, the proposals are split into *good-better-best* options. She will print out the proposals and grab pages of marketing material for the next meeting.

Day 45: Two More Meetings

With a book of collateral stuffed into her bag, the sales rep heads to the next two hour meeting. She offers a quick summary of what she heard last time.

Now comes a tricky part. The sales rep must present two or three solutions that fit the prospect. Each of these solutions includes concepts that are complex for the typical consumer. The rep spends time scribbling the harder ideas out on paper. She tries to show and simplify the concepts. If she is successful, the chances of closing the sale increase substantially.

The prospect looks over the mountain of materials. The rep answers any questions. If everything makes sense, the prospect fills out the application for the policy of his or her choice.

With the sale nearly completed, the rep heads back to the office and enters the customer's information into the system.

Problems plague this part of the process:

- Data-entry errors are common. Sales reps are human, after all.
- Customers often miss a signature or social security number line, and reps have to follow up by phone or fax. This kind

of error doesn't just add more time to closing the sale; it also results in lost sales.

When everything is ready, she sends the policy to underwriting.

Day 75: Closing the Sale

Eventually, the rep gets the policy back from underwriting. Once again, she makes a visit to the customer to make sure everything is clear. She hands over the gigantic policy document. The customer shoves it in a drawer somewhere and forgets about it for the next ten years. The rep calls once a year to find out if the customer wants to get together to reevaluate insurance options.

The rep completes the sales cycle—and what a process it is! Success is hard won for a Protectus sales rep. Those that achieve it understand how to work the system. With the right tools, success would come more frequently. What kind of impact could those tools have on Protectus as an organization?

Afterthought: Important Average Stats

Knowing she'll need numbers for future calculations, Sam decides to take a look at the results of the sales process. First she wants to find what percentage of leads in the initial pool converts into customers.

On average, 75 percent of leads are interested enough in the product to schedule a second meeting. The other 25 percent just don't have a need for insurance or aren't interested in working with the salesperson or the brand. They decline the follow-up meeting.

The sales rep, on average, loses another 25 percent of that 75 percent of leads willing to take the next step. These leads typically cancel a follow-up meeting and evade attempts to reschedule, eventually going cold.

We can express the percentage of leads that make it to the second meeting as:

$$75\% \times 75\% = 56.25\% \text{ of leads make it to the second meeting}$$

So, we find that 56.25 percent of the initial lead pool makes it to the second meeting. On average, 20 percent of those remaining leads will convert into customers.

We can see the total conversion percentage as:

$$56.25\% \times 20\% = 11.25\%$$

Roughly 11 percent of the initial leads convert to sales, directly leading to $5 billion in annual revenue. To put this in perspective, Protectus sales reps leave more than $45 billion of opportunities on the table with the other 89 percent.

It's unrealistic to be able to claim all of this revenue, no matter how efficient your processes become. But the possibilities are huge. We'll see the kind of impact the right suite of apps can have for Protectus later.

Before Sam can figure out how to claim some of that revenue, she has to find out what's slowing the process down.

STEP 3: IDENTIFYING THE CHALLENGES

One by one, members of the sales team meet with Sam and communicate their pain points. Many of the daily challenges and areas where real improvement could be made match the challenges she experienced herself as a sales rep. She identifies five bottlenecks that stand in the way of meeting this year's sales goals:

1. Too many customer touch points
2. A major time investment
3. Constant interruptions to the sales process
4. Data-entry issues
5. Too many people involved

Too many customer touch points

Visiting with the customer three, four, or sometimes *five* times before closing the sale is the norm for the Protectus sales team. Sam and her sales reps believe there must be a way to reduce the number of meetings between the introduction and the sale.

A major time investment

Besides multiple meetings, sales reps spend hours upon hours in transit or entering data, finding answers to questions, running scenarios, choosing which policies to pitch, printing collateral, and completing other administrative tasks.

Sales reps need time in the office to create a customized solution for the prospect. And what if they forget to run a scenario that the customer asks about? Chances are good they'd have to invest more time in an additional meeting to cover the gap. Could there be a more efficient alternative to all of this?

Constant interruptions to the sales process

Splitting time over multiple meetings can cause hot prospects to cool off. Rather than scheduling multiple follow-up meetings, Sam desperately wishes it were possible to close the sale in a single meeting. Reducing interruptions to the sales process could increase the percentage of closed sales.

Data-entry issues

Sam's predecessor kept a close eye on data accuracy, finding errors with about 2 percent of orders. What happens when this number is reduced to nearly zero? Would it help if reps didn't wait until the end of the month to enter notes into the system? Furthermore, salespeople and other admins record data multiple times—on the notepad and in the organization's CRM. There has to be a better way.

Too many people involved

To customize product proposals, salespeople must consult with financial and product experts in the office. Sales reps may be knowledgeable, but they don't have the skills to make certain decisions that accurately match the potential customer's needs. But what if sales reps had the ability to make these decisions on the fly, without the intercession of an expert?

Together, these obstacles represent kinks in the hose. They limit the potential that Sam sees in her sales team. If one or many of these issues could be fixed, she believes the goals laid out by management will be easily met.

Now she'll need to get creative. How can apps help sales reps overcome these obstacles?

STEP 4: GENERATING APP IDEAS

Based on the business challenges she identified, Sam creates a list of practical apps that could help Protectus improve the sales process, making communication and data collection much simpler.

She must consider her staff, ensuring a simple, intuitive experience. Each app concept should be simple for nontechnical staff to manipulate and address one or more of the issues she identified earlier.

App Concepts for Protectus

Company overview	Scenario modeling
Customer brochures	E-mail templates
Financial education	Order entry interface
Sales training	Document library
Customer management	Data collection and analysis
Sales rep dashboard	Customer engagement
Sales ranking	Training

A company overview

From the very beginning of the sales process, Sam sees a more compelling way to engage the prospect. A company overview app would feature a multimedia presentation on the brand, offering the prospect a guided tour through the Protectus experience. In this scenario, the sales rep hands the tablet to the prospect during the first meeting, guiding him or her through company background, videos, and other materials that detail the Protectus brand and its value proposition.

Customer brochures

Paper brochures have been a staple of sales rep visits for a long time, detailing specific products in the Protectus portfolio. Sam wonders if creating a customer brochure app could make the product education process more interactive. Considering the overwhelming number of brochures the organization creates, this app could also save money on paper and printing.

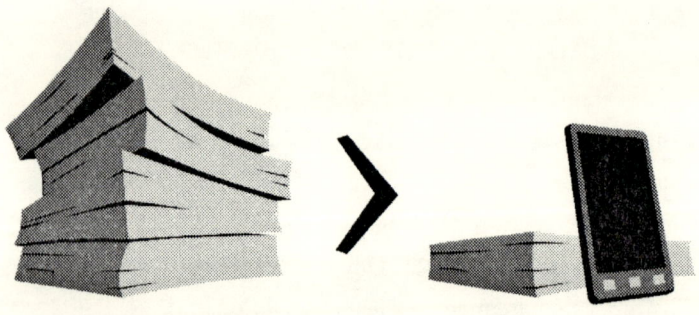

Financial education

This app would focus on explaining some of the more complex concepts that are tough for reps to communicate verbally. It would show how different financial products work and why Protectus customers should care.

Sales training

Training new sales reps takes time out of busy managers' days. Besides, reps are usually away from their desks, so traditional desktop training programs aren't very effective. A mobile training program could help them fill small windows of downtime with important role knowledge. Sam sees this app concept as an opportunity to save time as well as to create a more effective and interactive training program.

Customer management

Putting an easy-to-use face on the company's CRM could solve a host of problems. This app would mirror the sales rep's actual data-entry process before sending it to the backend system in real time. It would be a tool that reps use during the actual interaction with the customer, eliminating the need for double data entry and ensuring more accurate data collection.

Sales rep dashboard

The progress dashboard app would create a visual way for sales reps to monitor their own progress. It would include data like personal goals, monthly progress, upcoming appointments, and the like. Sales reps get a quick and simple way to check on how they're doing.

Sales ranking

Gamification could be a fun way to encourage a little friendly competition between the sales reps. Creating a sales ranking app would put top reps on a leader board, inspiring the rest of the sales team members to add more to their own tallies. For the organization, the app concept helps rally the team around the metrics that matter while making performance more transparent.

Scenario modeling

This app concept empowers the salesperson or customer to touch elements or change numbers of various financial offerings in real time to see what the effect is. Through interaction and visual modeling, customers understand better whatever they touch.

E-mail templates

Configuring e-mail templates for outbound e-mail can get messy. An e-mail template app would help make branded communications uniform and professional. It would help sales reps automatically link e-mails to customer entries in the CRM, giving them an easy way to review their previous communication with a given person.

Order entry interface

The order entry app ties everything together. Using the customer data and the scenarios, it creates a final order to suit the customer. This app actually records the sale and feeds it into the system, complete with the ability to save the customer's signature. After that, the rep's job is done.

A document library

A shortcut to sales collateral and other documents would help cut time usually spent retrieving the materials. This app concept requires a more capable search function that can scour the CRM's asset database.

Each of these represents a concept for a single app that can help the sales team do its job more effectively. Before she can calculate the impact, Sam organizes the concepts into suites of related apps:

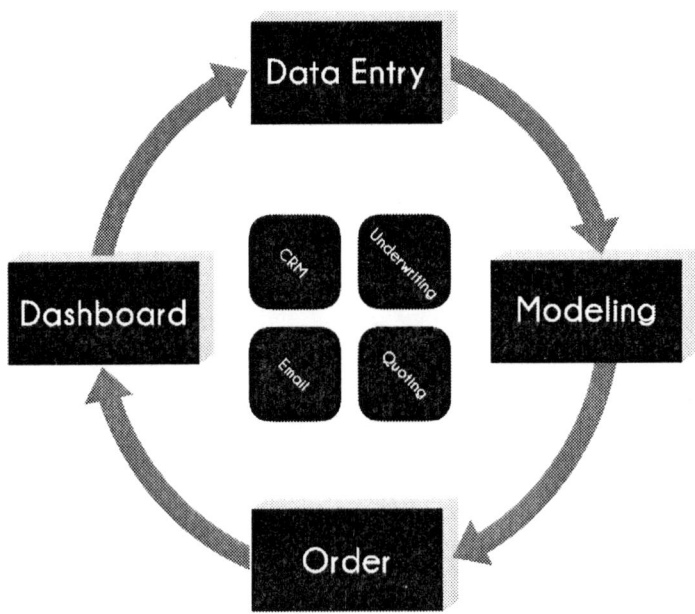

Data collection and analysis

The suite of data collection and analysis apps work closely with Protectus' CRM. This bundle of apps consists of Sam's **data entry**, **scenario modeling**, and **order entry** concepts.

Customer engagement

Here's a suite of apps focused on engaging customers to help them understand why they might want insurance and what's best for them. This bundle includes the **company overview**, **customer brochure**, and **financial education** app concepts.

Mobile training

Getting sales reps up to speed is tough, so the next suite of apps focuses on training and accessing information while in the field. The bundle features **sales training**, **e-mail template**, and **document library** concepts.

Sales ranking

Gamification and progress tracking apps could be major sources of inspiration for sales reps if they were packaged together in a suite. This bundle includes the **sales ranking** and **progress dashboard** apps.

These app bundles have the potential to connect the entire sales process from end to end in a way that Protectus employees have never dreamed of. Sam needs to do some math to find out which are most valuable.

STEP 5: CALCULATING THE IMPACT OF APP CONCEPTS

Management wants the numbers from Sam to back up the app concepts. The next step of the process is calculating the **return on app** (ROA).

The ROA is a number that describes the boost to top-line revenue or bottom-line savings the company stands to gain from an app concept.

To calculate ROA, Sam first needs to find the **new revenue potential** (NRP). How much more revenue could sales reps generate if they had, say, twice as much time to pitch?

Next, she must find the **cost savings potential** (CSP). The CSP describes how Sam's sales team will become more efficient. How much money can you save on sales reps as resources if you cut the time it takes to complete their workflows in half?

Finally, the BDA process includes consideration of the **organizational impact** (OI). The OI takes into account additional benefits that may result from making the workflow more efficient. The factors that go into this number are usually unique to the organization and the workflow. (OI can also include the "wow factor," or the impression that cutting-edge technology and efficiency can have on customers and staff.) Could the right apps reduce employee turnover? Do they put a progressive face on the brand? The numbers involved here need ballparking, but they are findable.

NRP and CSP calculations depend on four factors:

Impact (I)

The impact is the percentage of time the new process is expected to save in the typical sales process (as a workflow). If a change will save two hours of an eight-hour process, it counts as a 25 percent impact. The process improvement impact should still be estimated as a percent even if it can't be easily measured in specific times.

Time (T)

This measurement represents the percentage of their time that salespeople spend on the workflow in question. The time value is 100 percent if a workflow is all that a person does. If he or she spends four hours out of forty per week on it, the time value is 10 percent.

Revenue (R)

This is the average annual revenue value that results from the workflow in question. The entire revenue number should be used if this workflow generates all of the sales for the organization. If it relates to the execution of a subset of the business, the portion of revenue that it drives should be used.

Cost (C)

The total workflow includes many elements such as labor, materials, or any other hard costs. Labor costs represent a big opportunity for workflow revision to make an impact. We should also take into account all costs related to a process, like printing brochures, gas for transportation, shipping, rental equipment, and maintenance. It also includes other workflows that the workflow in question slows down.

To calculate the NRP, Sam uses the following equation:

$$I \times T \times R / (100\% - I)$$

This formula shows the additional revenue available if those on the team are able to do more with the time they already have.

To calculate the CSP, Sam uses this formula:

$$I \times T \times C$$

The CSP is a measure of potential labor or resource savings through adding process efficiencies.

The ROA is composed of the NRP, CSP, and OI. With these numbers, Sam can calculate an expectation for each app or app suite—its potential worth based on real numbers. We'll demonstrate how this works using the *Data Collection and Analysis* suite of apps.

Data Collection and Analysis Suite

Sam believes that simplifying processes with this group of apps could potentially cut the second meeting out of the sales workflow by integrating the research process into the first meeting. It returns customized insurance plans based on the scenario-modeling function.

This suite of apps is a window to the CRM and order-entry system. It eliminates double data entry by enabling the sales rep to enter data as he or she receives it. It saves the time usually spent putting in face time with financial experts and running different scenarios by automating scenario modeling in real time. Finally, it empowers the rep to make the sale on the spot and get the new customer registration process moving along.

First, Sam wants to calculate the NRP and the CSP. To accomplish this, she estimates each of the factors in the equation:

- Impact (I). With this suite of apps, Sam believes that Protectus could cut this workflow by 30 percent.

- Time (T). The workflow of converting a lead to a customer represents about 50 percent of the entire sales process.
- Revenue (R). The sales team generates the company's full annual revenue of $5 billion.
- Cost (C). The average sales rep's salary is $50K. Multiply that by a thousand salespeople to get a labor cost of $50 million.

Sam solves for the NRP:

$$I \times T \times R / (100\% - I)$$
$$30\% \times 50\% \times \$5B / (100\% - 30\%)$$
$$= \$1.07B$$

Sam solves for the CSP:

$$I \times T \times C$$
$$30\% \times 50\% \times \$50MM$$
$$= \$7.5MM \text{ in hard cost savings}$$

Based on the NRP and CSP, Sam estimates that **the ROA is between $7.5MM and $1.07B**. This suite of apps becomes a clear frontrunner in the BDA process.

Sam sees that these numbers don't fully represent the revenue potential of this app suite. She decides to calculate the OI to see what other benefits this app suite might have.

As we saw earlier, the OI represents tangible impacts of the apps outside of the pure process efficiencies. In Sam's case, it could boost the close rate by eliminating multiple meetings and converting the customers who go cold over the course of them.

Based on her earlier calculations, Sam knows that sales reps close 11.25 percent of their initial leads. She makes a conservative estimate that eliminating a second meeting could retain 10 percent of the leads that go cold between the first meeting and the second. That would mean retaining more of the 75 percent of the initial pool who scheduled that second meeting. The number lost would be 25 percent x 90 percent, or 22.5 percent.

Now, she can assume that:

$$75\% \times (100\% - 22.5\% \text{ going cold}) = 58.88\%$$

of the initial leads stay viable instead of 56.25 percent because she lowered the number of people going cold between meetings.

The new conversion rate is:

$$58.88\% \text{ interested leads} \times 20\% \text{ closable leads} = 11.77\% \text{ close rate.}$$

Protectus' sales team would now have an **11.77 percent close rate rather than 11.25 percent**. The resulting 0.52 percent improvement on the close rate represents a 4.6 percent hike in annual revenue. With this suite of apps, Protectus can potentially add another **$230 million** to their annual $5 billion in revenue.

All in all, then, the *Data Collection and Analysis* suite of apps could have an impact including $7.5 million in cost savings, $1.07 billion of new revenue from team process efficiencies, and another $230 million in additional revenue from a higher close rate.

STEP 6: SCORING THE APP CONCEPTS

Sam's next step is scoring the concepts and choosing the most valuable apps. The numbers she generates here will go a long way in helping her build a business case.

This part of the process underscores a crucial theme of this book: it's not about picking the app you *think* will work best for your organization. It's about choosing the *right* app for the job—the app with the biggest impact on the bottom line and the sales team.

For the Protectus team, it's also about priorities: the apps that perform best in the scoring exercise should be built first.

The first step of app scoring is building out a list of criteria like the one below. Companies like Protectus each bring unique needs to the table, so it's important to remember that the list is customizable.

Here are some criteria Sam used to rank some general costs and benefits of creating each app suite:

- How well will the app suite support our strategic goals?
- How much time will it take to realize the return?
- How much will it cost to create and implement the apps?
- How much time will it take to create and implement the apps?
- What kind of retraining will the new apps require?
- What percentage of workflow issues does the app suite address?
- What kind of risk is involved?

- Do we already have mobile-friendly programming interfaces or APIs built into our backend systems?
- How excited will sales reps be to use the new app suite?

To each of the answers, Sam assigns a score from 1 to 5 (with 5 as most positive) and tallies them up for a set of comparative App Success Scores.

Let's demonstrate how the process goes with the *Data Collection and Analysis* suite of apps.

- Sam believes the suite will have a big impact on the strategic goals, scoring it a (5).
- She thinks it should take very little time to realize the return (5).
- It will take a large investment to create this suite of apps (2).
- It'll take an estimated five to six months to develop and implement the app (3).
- The new suite of apps requires very little retraining (5).
- The app addresses 50 percent of existing workflow issues (5).
- Sam assumes a slight risk based on the cost and the amount of time to implement (3).
- Protectus' systems currently have very little in the way of mobile-friendly programming interfaces (1).
- After talking to a few members of her team, Sam believes the sales reps will be very excited to use this suite of apps (5).

The final score on this suite of apps is 34. It ranks first compared to the other apps Sam scored.

Now Sam has a clear choice for app development. She has a prioritized framework for attacking the development process. She's built a strong business case. It's time to get buy-in.

GETTING BUY-IN

Armed with a detailed presentation, Sam brings the case to the VP of sales. Her boss reviews everything over the weekend, adds her recommendations, and reports back to Sam that it was a job well done.

Later that month, Sam and the vice president team up to present the plan to the executive team. The comprehensive business case and the phenomenal impact it'll have on the bottom line are enough for the leadership to enthusiastically green-light the initiative.

As the project gets underway, Sam rests easy, knowing that she's reasoned this out thoroughly by following through on a proven process. With projections to back them up, app development has benchmarks already laid out. Her analytical decision-making will go a long way toward the success of the project.

Samantha Lee, Protectus' director of regional sales, is a successful change agent and a hero to the organization. She's set a precedent in her company that could help define Protectus' course for years to come. Sam has demonstrated that technology acquisition starts with process improvement.

As a workflow owner, she took workflow revision into her own hands. It's the impetus of new internal policy for Protectus—

that technology is the means, not the end. And Sam was the person who made that change happen.

Like Sam, do you have what it takes to be a change agent?

CHAPTER 4
What Is a Mobile Change Agent?

Ruairi Breathnach's CFO handed him an opportunity. As CIO, Ruairi used it to make the company a better place for everyone.

Oldcastle, a materials manufacturing company, maintains a unique business. A few years back, the company had built such a powerful presence in Ireland that it wanted to expand to North America. An outside consultant told company managers the move would be a mistake. Oldcastle ignored the advice and grew into the largest building materials company in the United States.

Much of that success is due to company acquisitions. Each line of business retained its autonomy, making for a decentralized organization. And with this structure, the company developed some internal needs. "You have a guy to grow the glass business and a guy to grow the asphalt business," Ruairi notes. "Eventually, you get to the point where they can benefit from more communication."

Oldcastle's upper management takes a proactive approach to business improvement, and recently recognized centralization of

IT as a major step forward. Naturally, management approached Ruairi for ideas.

The first step was the development of Infield, an internal platform for foremen to capture progress on projects. The application runs on laptops. For some of the foremen, these were the first PCs they'd ever owned.

Ruairi saw the platform's challenges immediately.

Non-intuitive technology

A learning curve prevented some foremen from getting up-and-running right away.

Wi-Fi access required

To access Infield, the foreman's laptop had to be online. Foremen would often have to search out a nearby coffee shop after work hours to fill out information.

Safety concerns

While a foreman could access Infield onsite if there was Wi-Fi, a clunky laptop was distracting from the most important responsibility: ensuring the safety of the team. Simplifying data entry in general did help alleviate this concern, but Ruairi knew it could be taken a step further.

Expensive technology

Each laptop cost the company as much as $1,200. Equipping an enterprise with laptops at that price point meant a formidable cost. Factor in antivirus and other necessary software, and the price rises.

Missed collaboration opportunity

Construction companies often collaborate based on project needs. "When you work with other vendors, it helps to create a collaborative environment," Ruairi says. "It saves a lot of time and hassle when it comes to communication." On its own, Infield did little to address this opportunity.

An accessible environment was the clear next step. Ruairi engaged an outside development team to create a mobile portal for Infield. He worked hard to get buy-in for his project. In his mind, this is only the beginning.

As a change agent in the IT department, Ruairi goes to great lengths to understand the challenges of workers and managers across all units of the organization and all working parts of an enterprise. Leading the charge for IT centralization means working with his disadvantages as well as his advantages.

WHO ARE CHANGE AGENTS?

People are the heart of all successful technology implementations. Enterprise mobility needs a human touch too. Visionaries like Sam and Ruairi spark tech innovation inside of organizations just like yours every day.

Redefining workflows and developing the corresponding apps requires know-how from various members of your team (and outside teams). A successful project assumes a squad of savvy professionals—people who know your business inside and out.

More important, enterprise mobility needs an origin. It requires an innovator with perspective, someone who understands the pain points clogging your organization's basic daily workflows. Only a practical, motivated person with the skills and clout to get things done can unlock your Billion Dollar Apps. That person is you.

Let's retrofit your old job description with a new title: change agent. A change agent is a visionary inside an organization who has high professional aspirations and the desire to make the company a better place. This person is a seasoned professional, soaring upward but still close enough to his or her roots to understand the needs of those being managed.

A change agent is an innovator who gets that technology is always going to be a big driver for the business. He or she embraces technology and, though perhaps not a "geek," scouts for new technology solutions or ways to apply technology that provide real business benefits. It's not just about the gadget factor of carrying around the latest device. It's pushing the limits to see where something new can make an impact.

Change agents understand that the bottom line starts with engaging the people on the front lines. They work to enact technology-driven change that will make lasting impacts at the core of their organizations.

Billion Dollar Apps are zero-dollar pipe dreams without a change agent to lead the charge. Without *you*, your business can only watch as the competition pulls ahead and out of reach.

WHAT MAKES A CHANGE AGENT?

Change agents are special people within an organization. They've typically spent less than twenty-four months in their current roles. In many cases, change agents manage their own teams.

Change agents have achieved progress before. Past projects may have been smaller, effecting technological change across a team or perhaps smaller changes across a department. Sparking enterprise mobility is a big step forward for a change agent, but one he or she has been building up to for an entire career.

Any department may have its own change agent. Here are nine qualities that change agents share:

Has solid professional experience

To be a change agent, you don't have to be at the pinnacle of your career. But experience is necessary. Change agents need enough perspective to marry workflow details with high-level goals. That experience may come from within the industry, but in some cases, change agents may be recruited from other industries to bring a fresh perspective.

Faces challenges head-on

Implementing enterprise mobility is no small task. It requires the type of person who has tackled challenges before—not just in the realm of technology, but also in general business issues. Change agents are invigorated by the challenges of workflow revision.

Comfortable with technology

You don't have to be a tech expert to be a change agent. But you do need to be comfortable with technology in general, especially with mobile devices like smartphones and tablets. Change agents also have an interest in how others are applying technology and aren't afraid to tinker with new tools.

Keeps pace with new tech developments

A change agent is up to speed on how technology is changing the world, both for businesses and consumers. Again, this doesn't require you to have a development or engineering background—just that you're tech savvy and studying the trends.

Draws parallels with other industries

Some industries mature more quickly than others. A change agent doesn't say, "Just because it worked for that industry doesn't mean it'll work for mine." Instead, he or she tries to find the common ground where the organization can borrow ideas to excel.

Listens to other people's challenges

Change agents see beyond the scope of their daily challenges. They know that many of their challenges hinge on those of their colleagues. Finding solutions to the problems that others experience will lead them to easing their own burdens.

Cares about the organization's success

A change agent is invested in the success of the organization. The time and effort spent at work every day has more than just a financial payoff—it's an investment that makes him or her empathetic with the organization itself.

Attains personal fulfillment through team success
Your career walks hand in hand with your organization. Change agents are highly motivated to find personal fulfillment in their work. They look at every opportunity to improve the organization as a chance to gain experience.

Mixes inventiveness and analytical thinking
Creative problem solving is a highly valued skill in a change agent. But it's also crucial to be logical and analytical throughout the process, using math and data to back up assertions.

While these are desirable traits for any professional, they're especially important for change agents. It takes a driven person to identify the challenges and find creative ways to overcome them through mobility.

OVERCOMING THE CHALLENGES

Change agents have foresight. They understand how business units and departments work together. Seeing the challenges that plague the organization is why you enact change in the first place.

Enterprise mobility tends to solve a handful of high-level business problems, no matter the department. These are challenges that slow down typical workflows across the enterprise. They've been solved before by other forward-thinking organizations. The problems your business experiences may be unique—but they aren't without precedent.

Let's look at a few top-level challenges that just about every organization has to deal with. Every single one of these can be eased through workflow revision and mobility.

COMMUNICATE WITH PEOPLE AND SEE WHERE THEY ARE IN REAL TIME

To stay competitive, business must occur in real time. What happens when you don't have the information you need at the exact moment you need it? This is a challenge that your smartphone has helped you overcome. But what about the rest of your organization?

Managers should know where service specialists (truck drivers, repairers, auditors, salespeople, etc.) are so they can make decisions on the fly. Service specialists should be able to open an immediate line of communication to support teams, including call centers, financial specialists, and account managers. Support teams should be able to report closed issues to managers. All this must be done in real time.

You can take the concept much, much deeper. Consider how much simpler logistics becomes when managers can see trucks in real time and communicate with their drivers.

What about salespeople in the field? Shouldn't they have always-on connections between their devices and CRMs? Why wouldn't you want a sales rep to close the deal on the spot? Does the time that typically passes between a verbal commitment and a closed contract erode the buyer's intent?

These are questions that some organizations don't examine very closely. But real time business—the ability to accomplish workflows in just seconds—is a critical part of how your organization will remain competitive over the coming decade.

Change agents understand this. They use their iPads to purchase goods on demand through Amazon. They answer important e-mail in the cab on the way to the next meeting. They're instant messaging managers from their iPhones when someone at a conference asks a question to which they don't have the answer. They access important documents through cloud services like Box and Dropbox.

Transactions can't happen in real time without mobility. Immediate access to reference and recording tools is the cornerstone of mobile business.

EDUCATE NEW AND CURRENT EMPLOYEES

Staff churn is a fact of business. So is the education of new employees. Both are frustrating and drain resources.

Education doesn't stop the moment an employee completes training, either. Many executives I've met over my time in business have emphasized that they're still learning, even after they've reached the pinnacles of their industries.

As a digital library the likes of which humanity has never seen before, the web has set an undeniable precedent in the world of education. We've yet to fully realize its potential as a teaching tool—although it's sure found legs as an instrument for passive learning.

Some of us have created tools for training and knowledge sharing, but some of them have failed to make an impact. It's time to rethink our approach in these cases. In other cases, the tools have been incredibly successful. It's time to take those to the next level. We must make them mobile.

Satisfaction with current tools isn't part of the change agent's job description. You must imagine how you can take digital education and training further. Organizations like yours accomplish this by:

Feeding trainees digestible bites of education

Training apps simplify how mobile workers take advantage of short time windows between events. Workers can absorb quick slices of information pretty easily. Mobility can make learning visual too, through short videos designed to fill ten-to-twenty minute slots.

Leveraging tools with lower learning curves

As we've demonstrated, intuitive devices like smartphones and tablets are simple to use. Mobile apps that rethink workflows make them even easier. Our process for mobility can greatly reduce the training burden, making every process that much easier to accomplish.

Making knowledge more accessible

Digital storage made information searchable and therefore easier to reference. Mobility makes recorded knowledge even quicker to access. Reimagining workflows can take some of the burden off of the worker too, by guiding him or her to a swifter, more accurate resolution.

Creating training programs that are more engaging

Mobility represents an opportunity to make your training more engaging. Coupled with simplicity and accessibility, the possibilities for a more knowledgeable workforce are astonishing.

Protectus had a simple need: an empowered sales staff. The company wanted its mobile sales force to have the resources they needed at their fingertips. No more trips back to the office. No more scrambling to schedule second and third meetings. The sales team needed to pull up information quickly. Mobility delivered a solution. Accessing information on the fly is a huge benefit of mobility.

FIND WAYS TO ANNIHILATE BOUNDARIES TO COLLABORATION

Interdepartmental collaboration is another problem with which we've grown complacent, and yet we still feel the pain of the bridges missing from our communication. Every day we try to make it easier to talk over the walls that separate departments.

Silos keep the enterprise organized, but that doesn't mean they can't work together. Where's the middle ground?

Understanding the motivations of other departments is an essential part of being a change agent. It's a challenge that change agents should want to tackle head on. Not just for themselves, but for the good of their colleagues.

Distance also hinders collaboration. In this day and age, why is this still the case? I've seen a number of organizations struggle with regional offices built around "tribal knowledge." Each office may share information internally, but those best practices and good ideas don't make it to the rest of the company. Lots of effort is wasted in recreating solutions that already exist and grappling with problems that have already been solved.

Mobility bridges the information gap by empowering communication. This takes a variety of forms—a mobile-enabled enterprise social network, a more comprehensive shared knowledge base, data from one business unit quickly shared with a different business unit, and so on. It tears down geographic walls by enabling people to collaborate and share information in real time. Lowe's offers a simple example of how this works: if a part isn't in stock, floor personnel use the mobile app to find and reserve that part in a different local store.

MOTIVATE PEOPLE

How much better would your company be if people were excited to come to work every day? What about if people were consistently tuned in to their own performance metrics? This is the sort of real-time feedback the millennial generation craves and expects.

Mobility provides an opportunity to rethink how employees tune in to the enterprise. Billion Dollar Apps can be motivators in three very distinct but important ways:

Simpler workflows and devices translate to happier, more motivated people.

The very act of simplifying the lives of employees is a huge step toward a more motivated workforce. People who don't dread coming to work every day tend to be more motivated, after all.

Investing in people motivates them.

As an employee, seeing your employer make a direct investment in tools to help you motivates you to do better. It helps instill a sense of loyalty and the feeling that managers and executives care about the people they direct.

Mobility offers the potential to gamify productivity.

Gamification is a huge opportunity across the world of business. Just giving workers a glimpse into how their goal accomplishments and periodic milestones match up against their coworkers' can add friendly motivation. (Implementing gamification well takes the right touch, of course.)

As a change agent, you're a self-motivated person. You may even have your own system for measuring goals and recording milestones. But not everyone across the enterprise has an inherent drive to succeed. Mobility is an opportunity to change that. Gamification is a great way to rethink your workflows, especially if you don't already have a goal-tracking system in place. It's the practice of presenting bite-sized goals and metrics to someone to motivate him or her to take action.

Think of recent gamified consumer activities. FitBit, for instance, makes exercise more interactive by tracking fitness and diet goals. On a smaller level, social sites like LinkedIn gamify profile completion by showing your percentage of progress.

In sales, a little friendly competition is a typical motivator. Giving sales reps an always-on leader board is a great way to incentivize better production.

Through mobility, you get an opportunity to simultaneously add dashboards and make them more accessible for people, no matter where they are or what they're doing. It's a great way to incentivize data collection too, giving your organization more insights into how it conducts its daily business.

PRINT, CARRY, AND FILE PAPER DOCUMENTATION

Who uses paper these days? More organizations do than ought to. Paper is still a burden for a lot of companies. It's still the standard for a variety of industries.

Organizations that still use paper run into obstacles. Among them are double data entry and general data errors (which we'll discuss shortly). Workflow inefficiencies like filing, stapling, printing, copying, scanning, and more are major hang-ups throughout paper processes.

Beside the environmental concerns, the cost of paper and ink is so far beyond the scope of electronic documentation costs that an organization could save millions of dollars a year just by eliminating them. Let's not forget the implications that relying on paper has for field specialists, salespeople, and remote workers in general: massive, disorganized briefcases and counterproductive trips back to the office just to get the right materials.

Is any of this necessary? In a few years, it won't even be an *option*. The way consumer trends are moving, it won't be long before the phrase "some customers prefer paper" is wiped clean from management's excuse handbook. It's time to start phasing it out of your daily business.

Some forward-thinking restaurants are equipping servers with tablets. They're saving paper and decreasing the chances of mistakes made when transferring orders from a notepad to the system. They're also processing payment for the check at the table, making the experience quicker and more secure for customers, who don't have to watch their cards disappear into a back room.

We just watched Protectus take such measures. Sales reps no longer have to print out books and books worth of collateral to bring to visits with prospects. The cost and time savings are enormous. Cutting paper and ink out of the day-to-day routine is a necessity for everyone. As a change agent, you have the power to get it done.

USE AND MAINTAIN DEDICATED SINGLE-USE HARDWARE

Bar code scanners, RFID readers, and cash registers are quickly going the way of the luggable. Why would you carry a laptop from site to site? Why would you lock an employee behind a cash register when he or she could be walking around helping customers?

As a change agent, it's your job to think, "What if?" Mobility can potentially replace any piece of hardware associated with accomplishing a workflow. This is especially crucial for devices that were only created to perform a single task.

I once worked with a large brand that had thousands of field specialists across the globe. Each specialist carried a pair of devices around on the job. Each device accomplished a specific task.

Forget the problems this causes on the enterprise level for a second. Just imagine the frustration that field specialists went through on a daily basis.

We saw two major issues with the situation:

A big, fat waste of time.

The devices needed a specific connection to push the data into the system. Therefore, specialists were unable to enter information on the job; instead, they performed data entry for hours after fieldwork. In many cases, they went through the motions without paying attention, making mistakes that they had to go back and fix. As a result, specialists worked more than a dozen hours a day...even on a good day!

Broken devices.

The frequency of broken devices was a major cost for the organization. It's not that big of a surprise that making specialists responsible for more than one device ended up in negligence. Imagining that workers *wanted* their devices broken isn't even that big of a stretch. (Have you ever wanted to throw a frozen PC against a wall? It was a lot like that.)

A new breed of mobile connected devices are emerging that take advantage of our mobile devices for their brains and their user interfaces. The user still interacts with the phone but takes advantage of the capabilities of printers, bar code scanners, medical devices, or other specialized sensors. These connected devices, paired with phones or tablets, will replace the specialized, single-use hardware of the past and provide much broader functionality.

A great example is the cash register or point-of-sale system that is rapidly disappearing. iPad-based POS systems not only process a transaction anywhere in the store, but because of the apps running them, they give the sales associate more information than ever before to help with customer interactions.

ENTER DATA IN A TIMELY AND COMPREHENSIVE FASHION

Most of us aren't data scientists. But as a change agent, you should understand that data has a significant impact on your organization. Collecting the right data helps enterprises accomplish a variety of tasks, including:

- Finding bottlenecks in processes (like the ones we're examining)
- Creating better products
- Predicting trends in the industry
- Developing stronger marketing programs
- Identifying market gaps

These are just a few examples. As analysis technology gets more advanced, you'll see it used more and more throughout the organization.

There are hundreds of thousands of viable scenarios where data makes a major difference in how an organization does business. For now, collecting data often relies on people. Sam's salespeople at Protectus input data into their CRM, for example. A field specialist at a pest control firm reports back on a location visit. A town car driver tells a mobile app that his car is empty, and he wants a fare. In each of these scenarios, someone is responsible for entering data. The tougher it is for that person to do that, the more likely it is you'll get incomplete, incorrect, or late information.

This is a problem that plagues businesses across the globe. What's the point of collecting data if it isn't accurate? As a change agent, how do you make data collection more effective for the organization *and* reduce the burden on the individual employee?

One of the major obstacles to more efficient data entry is that companies treat it as an afterthought. A big advantage of mobility is building data entry into the workflow so that it's part of the process. As a user looks up information, a well-designed system automatically records what he or she does and updates customer and sales records without the user having to take an extra step of making separate notes in the CRM.

Our friend Sam understood this challenge only too well. It's hopelessly interconnected with the proliferation of paper and extraneous hardware. With mobility, Protectus met the problem head on. The company was able to ease the burden on the sales staff and create a more efficient and productive business development process.

A more productive sales team meets goals much more easily. But workers at pre-mobile businesses have a much different perspective when it comes to measuring success.

DO MORE WITH LESS TO MEET HIGH EXPECTATIONS

At the heart of all of these challenges is the constant pressure to meet expectations, no matter how difficult that may be.

No one said your job was supposed to be easy. But there are such things as *unrealistic expectations*. Thousands of businesses enforce them every day.

Earlier, I mentioned an organization where field specialists work more than a dozen hours a day. There's no work–life balance there. The expectation that those specialists will meet their goals in a normal amount of time is unrealistic. But the business can't function without meeting those goals. So the workforce toils on for long hours to collect a paycheck.

If employees can get the job done faster, it's better for everyone involved. They shouldn't have to spend hours filling out electronic forms back in the office or on a laptop from home or a hotel. That kind of experience leads to mistakes and unhappy workers. The right tool is all employees need to meet those expectations, on time and without breaking their backs.

Based on everything we've seen so far, mobility is the clear solution to this challenge. It's the tool that makes everything easier across the enterprise. A stronger bottom line is the main goal. Happier, more productive people are the result of simpler workflows. The right mobile device is the tool. Billion Dollar Apps tie everything together.

Now that we have a high-level view of the change agent, it's time to dig into the Billion Dollar Apps process from top to bottom.

CHAPTER 5
Billion Dollar Apps Process Overview

Mobility is no small undertaking. It requires a planned approach. Charging ahead with mobility without that plan is destined to fail. That's where the BDA process comes in.

Like each element of business they address, Billion Dollar Apps are discovered through a process. I've used this process again and again to find mobile success for enterprise organizations. We've already seen it in action in the story of Protectus. Many of the real-life examples (including Oldcastle, Lowe's, and more) used this process or one much like it to rediscover millions of dollars in lost efficiency.

By implementing the BDA process, a change agent can:

- Tie process changes and technology acquisitions directly to top-level organizational goals
- Put high-level and granular business drivers into a connected and relatable context

- Understand where kinks in the hose occur for a specific line of business
- Find solutions to problems that have long plagued workflows
- Prioritize apps based on how they impact the organization using the App Success Score

The first part of the BDA process is dedicated to shining light upon the conditions under which your line of business operates. These first three steps are opportunities to gain a deep grasp on the bits and pieces that make up each of your workflows. Armed with this information, you have the opportunity to make the changes necessary to cut costs and increase revenue potential.

BDA Process

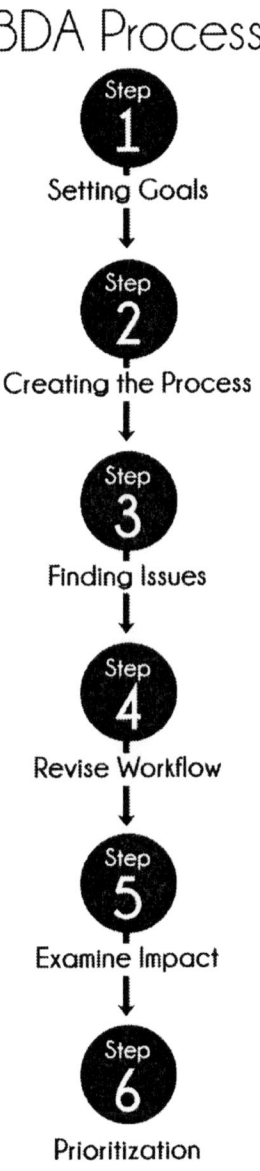

Step 1 — Setting Goals
Step 2 — Creating the Process
Step 3 — Finding Issues
Step 4 — Revise Workflow
Step 5 — Examine Impact
Step 6 — Prioritization

The steps we'll cover in the first half of the BDA process are:

1. Putting overarching goals in perspective.
Before you jump into your mobility project, you need business perspective. What is your organization trying to accomplish this year? What about over the next five years? The next decade?

2. Finding the workflows that form the process.
The second step brings us in for a close-up perspective. After you've seen things from the management level, it's time to really understand how your teams accomplish their daily goals.

3. Pinpointing where workflows go wrong.
Before you find the solutions, you should know exactly what the challenges are. In this step, you use your newfound perspective to understand why workflows aren't as efficient as they could be.

The second half of the BDA process is all about creating concepts to solve problems and projecting what will happen when you bring those concepts to life. It's an opportunity to take concepts down to numbers and validate your priorities.

Forming the predictive models you'll need to build a business case is exciting. You get a chance to envision your ideas and see real results on the page. The next three steps of the process are focused entirely on how to form ideas and understand the impact those ideas could have on your organization.

It's critical to note that you shouldn't just go through the motions on these steps, especially the final two where you'll start building a business case. The theme to remember here is **choosing the right apps, not the obvious ones**. Without the numbers, the assumed impact that obvious apps might have on the organization can be incredibly deceiving.

A.M. Castle and Co. is an example of an organization that did the legwork, ensuring it went down the path with the greatest return. In this case, the obvious app was a price-quoting tool for the outside sales team. But following the next three steps of the BDA process, the organization found concrete evidence that the obvious app wasn't worth anything to it in the long run.

These three critical steps are:

4. Nurturing ideas for workflow revision through mobility.

With workflow challenge top of mind, it's time to get to the drawing board. Here's where you'll get creative, finding ways to solve problems and create greater process efficiencies.

5. Calculating the impact of your concepts.

This step focuses on the math behind choosing the right concepts for your business. Here we calculate the potential impact that each app concept can have on the workflows in question.

6. Scoring and prioritizing your concepts.

Just before you get your mobility project underway, you take a look at factors outside of each app concept's impact—issues like timelines, worker adoption, and general feasibility. Combine the organizational impact with project conditions to prioritize development.

    The end result of this process, as you now know, is a prioritized list of apps that can be used as a roadmap for mobility in the organization. Each app will have financial metrics that can be used to measure success.

First, let's take the time to understand exactly what your organization needs to accomplish from the high level.

CHAPTER 6
Step One: Goals in Perspective

Laying the groundwork for your organization's mobility starts with envisioning a destination—an endgame congruent with your organization's goals.

Change agents are in a unique position. On one hand, they're not very far removed from the people that make up an organization's workforce. Executives and other managers may not be so lucky. On the other hand, they're also close to the goals declared by executives. They understand why those goals are important for the livelihood of the business and may even have some hand in shaping them.

Such goals might include:

- Boost revenue by a certain percent
- Cut costs and boost the bottom line by a certain percent
- Reduce customer churn by a certain percent

Usually executives put forth three to five high-level goals like these over the course of any given twelve months. The close perspective of the change agent helps him or her plan a framework for the first step in the organization's journey to becoming a mobile business.

To be successful, Billion Dollar Apps must focus on these organizational goals— benchmarks that drive business success and help your firm avoid pitfalls that lead to creating the wrong apps.

Building apps that aren't focused on these goals creates a much smaller impact. These are likely to be the "obvious apps." The results will be limited to a small subset of the organization rather than supporting the overall strategy.

The goals you pursue can be organization-wide or at a division level. Just be careful not to let the goals become too focused on the needs or issues of a small part of the organization, even if that part consists of members of the C-suite asking for the obvious apps.

Protectus had goals handed down by its executive team. To keep up with industry leaders, the company had to take an aggressive stance on growth. A 3 percent annual bump just wouldn't cut it anymore if the organization wanted to join the ranks of leading brand names. The insurance company identified the numbers it needed to reach to accomplish its qualitative goal of market growth: it had to grow top-line revenue by 10 percent that fiscal year. This goal was conditional, relying on sales to increase its close rate without adding personnel.

Protectus started with a clearly defined endgame—but no concrete pathways to achieve it. Your situation prior to a mobility project might be similar.

Write down the goals the company has set. Having them clearly documented is the first step on the path to Billion Dollar Apps.

WORKING WITH QUALITATIVE GOALS

Not every organization lays out very concrete goals. Benchmarks like those Protectus identified were some of the first of their twenty-year existence. Some organizations create qualitative goals like "surpass competitor A in brand awareness" or "make the Inc. 500." It's up to you to tie realistic numbers to these goals that you can then use to calculate the impact your app concepts have.

Identifying these softer metrics can be tricky. Here are a handful of ideas for finding numbers that might apply to your situation:

Track the competition.

Is "surpassing the competition" one of your organization's short-term goals? A competitive analysis could come in handy. Record competitor growth trends and compare them with your own. Come up with viable numbers your organization needs to reach to establish market dominance. Find ways your department can contribute to these numbers so you can give them context.

Analyze customer service.

Organizations that rely heavily on recurring revenue or return customers may target improvements in customer service and customer retention. Here you'll want to look at numbers like wait times, customer satisfaction, and closed tickets.

What's the value of each retained customer to the bottom line? They forego the high cost of customer acquisition for the much lower cost of retention. That bottom-line impact is another number to consider.

Take each of the numbers and decide on a percentage that represents a strong but realistic improvement. Try to tie these numbers to overall customer retention and the impact it could have on revenue.

Increase annual revenue.

Your organization's goal may be as simple as revenue growth. Like Protectus did, this means identifying how your department drives revenue and using numbers to quantify that impact.

Finding the sales department's impact on revenue is pretty straightforward. It may be less so for other departments. But different product lines or departments do have a major impact on revenue. What are the high-level growth targets for each business unit?

These are just a few examples of ways you might go about quantifying organizational goals. It's always a good idea to meet with executives to flesh out these numbers. With organizational goals in perspective, it's time to move on to the workflows that make your organization tick.

CHAPTER 7
Step Two: Finding the Workflows That Form the Process

Now, we move from the big picture to the working parts. How do the workers on the front lines of your business do their jobs? What tools do they use? What does a day in the life of a staff member in your department look like?

Metrics are incredibly important here. Using the right apps, what dials can you turn to drive efficiency within a given process?

To analyze workflows, you have to look at every element and interaction within each process. This step consists of exploring the daily tasks of your workforce, isolating the different processes that make up their responsibilities, and finding the processes that drive your organization to accomplish the goals outlined in BDA Step 1.

The workflows crucial to this step are only those that will have a direct impact on the goals you identified. If your goal is revenue generation, which workflows support it? (Sales workflows may be most relevant here.) If the goal is boosting the bottom line through cost cutting, you might focus on production and operational workflows. The key is not to just pick any workflow with issues. If you take that approach, you'll likely end up with obvious apps that don't deliver enough for the organization.

For Protectus' sales process, we identified the series of meetings that start with interacting in person with a lead and end with converting that lead into a customer. This process represents half of a salesperson's time. (The other half is dedicated to cold calling and lead generation.)

Change agents usually fill a role that requires them to ensure the machinery works efficiently. Here you need to act like an investigative journalist, finding creative ways to really understand how workers power the enterprise.

There's a common misconception that only manual, repetitive processes can be improved with mobility. I worked with one group that claimed their sales process was too intricate to improve in these ways. In actuality, they had no sales process. They relied on their salespeople to just go get the business. Every day, those salespeople took dozens of different approaches across the country. They had no standard process. Each office created its own processes and tools without sharing best practices. One or two of those approaches actually worked best across this organization. Finding and standardizing them might have paid huge dividends.

STRATEGIES THAT HELP CHANGE AGENTS GET CLOSER TO THE PROCESS

As change agent, it's your responsibility to step into your worker's shoes: to fully understand the process, you must get close to it. As I noted, this'll require some careful investigation.

Here are a few strategies. Used together, they offer you the visibility you need to isolate workflows for analysis and revision later on.

Shadowing

Shadowing is following a worker's daily routine. It may be as simple as following someone through the office. It also takes the form of a ride-along—a trip with field specialists, sales reps, installers, or other mobile workers moving from place to place throughout their workdays.

With shadowing, you can see firsthand what a worker accomplishes (and in some cases, what he or she doesn't accomplish). You can watch workers interact with the tools that either help them get there or stand in their way.

The worker's perspective is essential down the line. Take extensive notes on the workflows that make up the day. Try to understand the pain points the worker has. Take stock of everything. You can organize it more effectively later.

Workers may not even realize some of the inefficiencies you'll see. When you ask about them, you may hear the dreaded "It's the way we've always done it"—a great indication that something can be improved.

Interviews

While shadowing is focused on experiencing workflows, interviews are focused on what workers think of those workflows. You can couple an interview with shadowing or arrange a series of interviews to get a good sampling of different workers.

Ask team members what gets in their way and how they would fix the process. Some of the best ideas come from the front lines.

Below are some great starting questions to get to the root of the workflow.

- What are the three primary work activities of your department?
- What are the main steps of each activity, and who does them?
- How much time does it take to perform X?
- How many times per day (and week) do team members perform X?
- What information do you need to accomplish X?
- How do you get that information?
- Is there movement from place to place?
- Where do steps take longer or involve more people than they should?
- What task takes the most time to perform?
- What would you like to simplify about the process, if you could?

Data analysis

Data is at the core of every business system that makes up your enterprise. Most of it sits within reach but untouched. Today businesses have a responsibility to mine and analyze that data. With it, you get a completely different (and necessary) perspective on the details that make or break your processes.

Data analysis goes beyond software platforms. Enterprise organizations can analyze trends over time, identifying when and why productivity peaks or dips.

Today's data scientists are doing amazing things. But you don't have to be a scientist to understand what certain types of data are telling you. If you have access to it, use it. If you plan to boost revenue, look at data like:

- What's our revenue per business or product line?
- What's our customer retention rate?
- What's our conversion percentage between step A and step B?

(Take a look back at how Sam calculated Protectus' OI for an example of where data can come in handy.)

Equipment and tools

Mobility may introduce new tools and hardware into the enterprise. To build a business case, you need a grip on why legacy devices and luggables aren't as effective as they should be.

To understand the workflow, understand the tools that workers use to accomplish it. Using the equipment for yourself (if it doesn't require extensive training or certification) can help you get the perspective you need.

One group I worked with had a luggable data collection device that required users to start an entire workflow over if they had to go back and change data. That made absolutely no sense and cost them quite a bit of time in the field.

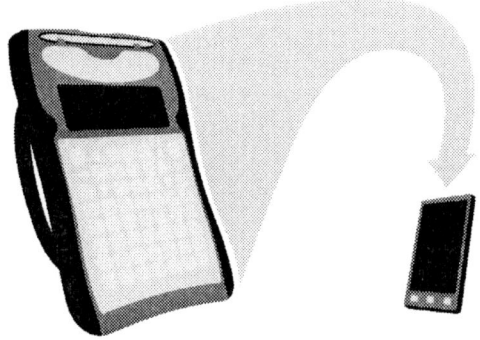

With an insider's perspective, your next step is to create a map of a worker's responsibilities. That way, you'll have an organized document to work from for the next step of the BDA process.

SEGMENTING AND ORGANIZING WORKFLOWS

Next, build a spreadsheet to get a full, organized view of your workflows. The spreadsheet will help you visualize the working parts so you can isolate inefficiencies later.

Here change agents must ask lots of questions and gain insight into the detailed workflows of the organization. It's not about having the minutiae mapped out— it's about digging for the biggest efficiency opportunities.

Look to answer questions like:

- How long do steps in the workflow take?
- How much of his or her time does the average worker spend doing that workflow, or how many times is it performed each week/month/year?
- What prevents the workflow from moving to the next step (like sales opportunities not converting from a lead to a closed sale)?
- Where does that happen, and why is progress blocked?

Your spreadsheet should include several different categories that you can fill out after you've had a chance to experience the process firsthand. Here's a step-by-step breakdown of how to work with the information you gather. Enter everything in your spreadsheet.

STEP 1. BREAK YOUR WORKER'S ROUTINE INTO STEPS.

Divide the major pieces of a worker's week into overarching categories—the main tasks that go into completing a process. Protectus analyzed the process of finding leads and converting them to customers. Sam broke out the sales rep's routine into these typical categories:

- Cold calling to set appointments
- Initial face-to-face meeting
- Researching and preparing proposal
- Second face-to-face meeting

STEP 2. RECORD HOW MANY TIMES A WORKER PERFORMS A WORKFLOW IN A GIVEN PERIOD.

Some workflows are repetitive. Others may only need to occur once or twice a month. Recording how often a worker typically performs these tasks is important for understanding time commitments and where major efficiency improvements might be made.

Protectus sales reps perform their initial face-to-face sales meetings twenty times a month. The second face-to-face meeting occurs about a dozen times over the same period. Cutting back on second meetings frees up time, serving to increase the number of first meetings and therefore to help Protectus improve its bottom line.

STEP 3. ESTIMATE HOW MUCH TIME EACH WORKFLOW TAKES.

Saving time means saving money and freeing up resources to drive more revenue. Understanding just how long each of these workflows takes will help you figure out where eliminating elements of workflows can make an impact on a large scale.

With BlueStar, we saw that energy auditors spent twelve hours to generate and submit a single field report. (It only takes four hours now.) A.M. Castle and Co.'s 96 percent reduction in research time began as a ninety-second workflow. (It only takes two seconds now.)

Look at the time a single workflow takes. Multiply that by the amount of times the workflow is performed to get a better overall picture of its time investment.

Pay attention to calendar time lags in a process as well. For Protectus, we saw about fifteen days pass between the initial meeting and the secondary meeting where the sales rep prepares the proposal. Is that calendar lag affecting the successful outcome of the process? Does it cause sales to stop or other workflow elements to cease because of the time delay?

STEP 4. SPLIT STEPS INTO THE ACTIONS THAT MAKE THEM UP.

Break down workflows further. Look at each of the actions that make up a given workflow. Split them up and categorize them under each of their associated workflows.

The workflow for Protectus' initial face-to-face meeting is split into four tasks:

- Explaining the company background and differentiating Protectus
- Discussing the fundamentals of insurance with the lead
- Gathering personal financial information and long-term goals from the lead
- Scheduling the next meeting

To get more perspective on each action, look to the people who make it happen.

STEP 5. NAME EVERYONE WHO TAKES PART IN EACH ACTION.

In our grain silo example, the process required two separate people to accomplish the task of filling the truck. With mobility, the organization was able to give full control to the driver, simplifying the workflow to the point where a single person could accomplish it. Following that logic, it shouldn't be difficult to see why you want to take stock of everyone involved in a process.

Look for workers in supporting roles—people preparing reports, consulting on projects, performing administrative tasks, etc. Mobility sometimes frees up collateral workers to focus on their main responsibilities, removing them from workflows owned by people in other roles.

For Protectus, most of the actions involve the sales rep and the customer. But other staff members (like financial planners and admins) are brought into the process at different points.

STEP 6. LOCATE WHERE EACH ACTION TAKES PLACE AND HOW MUCH TRAVEL IS INVOLVED.

Location and travel are major factors for some workflows, especially for sales reps and field specialists. Time spent in transit can take up a significant amount of a workflow's average time to completion.

This is an especially important concept for field service organizations. A "truck roll"— every time a service vehicle has to visit a customer site, especially unnecessarily —costs money. Remote diagnosis of problems or self-service capabilities for the customer can offset this.

Protectus sales reps have to travel back and forth for their meetings. The time they spend traveling is not insignificant to the process. On average, reps travel five to fifteen miles for each meeting. (Knowing what you know, you see where cutting one of those meetings out makes a huge difference.)

Based on that knowledge, estimate the cost of the travel. How much time is spent in transit that adds no value? What is the gas or vehicle cost? You'll need this information for Step 9.

STEP 7. ESTIMATE HOW MUCH TIME EACH ACTION TAKES.

Get more granular with your timing. You already understand how long each workflow takes. What about each individual action that makes up a workflow?

Record times during the shadowing process. Then discuss them to validate what's typical. Times may vary wildly depending on what happens on a given day. But the goal is to get to some averages and common cases. If you don't have much to go on, estimate the percentage.

In the second meeting, Protectus sales reps spend fifteen minutes just to reengage the customer! Right away, we can see where time could be used more effectively. That second meeting also includes more time spent running through the proposal and fielding questions the rep may already have answered in the first meeting.

STEP 8. ESTIMATE HOW OFTEN THE WORKFLOW IS PERFORMED SUCCESSFULLY.

What percentage of these workflows actually yields bottom-line success? In some cases, almost every workflow accomplishes its goal. In others, there may be a very low completion rate. Whatever the situation, this is a crucial number to have handy.

Take service organizations, for example. Return trips for additional service when a visit doesn't fix the problem can be very costly. What percentage of those first visits results in successful outcomes? What percentage requires return trips?

As we saw, Protectus found a 75 percent conversion rate on the initial meeting, followed by a 75 percent success rate on booking the next meeting after the proposal had been drafted. These numbers are critical in finding the lost revenue that Billion Dollar Apps can recover.

STEP 9. RECORD THE COSTS INVOLVED IN EACH ACTION.

The previous step of your worksheet involved recording any recurring costs outside of the worker's pay—materials, gas, consultants, etc. A full picture of costs will help you assign more accurate values to different app concepts later on.

Protectus sales reps spend twenty dollars every time they print materials for their second meeting. That meeting occurs twelve times a month, on average. Factor that in for thousands of sales reps and it becomes a major expenditure for the organization. (Don't forget the reimbursement costs for travel between appointments.)

As we saw in the CSP equation, Protectus also spends $50K on a thousand sales reps. This is a $50MM cost to take into account. Create estimates for any of the costs that are unknown in Step 2 and document the assumptions.

Your spreadsheet of all this data becomes an important tool for the next step of the BDA process.

CHAPTER 8
Step Three: Identifying Workflow Issues

We've already seen a slew of examples where today's technology can improve the efficiency of workflows. Figuring out where mobility can make things easier for your workforce is both art and science.

Step 3 of the BDA process starts by looking at the numbers you collected in Step 2 to help you understand which workflows have the most impact. The end result of this step is a list of issues that hinder workflows. Along with each issue, you'll attach a value that represents its impact on your organization.

First, change agents must get creative in figuring out where workflows could be better. Here's where it helps to ask the question, "What if?" The more you ask, the more opportunities you'll find to make processes quicker, simpler, and more effective.

This part of the process should be prioritized based on what you already know. Spending lots of time analyzing a workflow that is only a problem twice a year or for two hours isn't worth it. Instead, focus your time on frequently used workflows that affect the largest number of people.

Second, change agents must don their scientist caps. Assertions that result from asking "What if?" are testable hypotheses. You must do everything in your power to understand the impact that revising an action or workflow can have on a process. This means digging deep and doing the math.

Overall, workflow revision should follow an analytical approach:

- Start where workers spend the most time or incur the highest costs.
- Find problems that confront these workflows.
- Look at the process from end to end.
- Imagine what a workflow might look like if you cut out a step or removed one of those issues.

Keep in mind that areas where you boost efficiency may not come directly from mobility. Rather, remember that mobility should be used as a tool to force simplicity into your processes. This mindset can help you see a data-entry workflow as more than an opportunity to put a mobile face on a web-based system—it's a chance to simplify the steps necessary to accomplish data entry, and do it more comprehensively.

First, let's take a look at strategies that can get you closer to the workflows.

WHAT TO LOOK FOR WHEN PINPOINTING WORKFLOW ISSUES

Getting closer to the process is an important part of the BDA analysis. But it's equally important to know what you're looking for.

For health care, customer-facing apps are enterprise-grade, simplifying workflows for staff in extraordinary ways. Former PayFlex CTO Tony Dillon has a straightforward approach to this step of the BDA process. "We look at the friction in workflows," he explains. "Where are the greatest costs? Where is the greatest friction between our customers and us? Obviously, we want to remove as much of that friction as possible."

Tony spends his time looking at traditional points of contact with customers like interactive voice response (IVR) platforms and websites. As we saw earlier with PayFlex's story, claims processing is one of those billion-dollar opportunities in health care.

Tony's experience in mobility gives him a keen eye for areas of improvement. He typically looks in these areas for issues:

Friction

As Tony mentioned, anywhere you find friction across a given workflow represents an opportunity for improvement. Friction is usually obvious at customer touch points. It can be less obvious in other places across the enterprise.

Friction occurs when more than one workflow, tool, person, touch point, or other process element rub up against each other in a negative way. Protectus experienced friction in how sales reps interacted with their CRM platform, for example. The program just didn't work intuitively enough to make them care about how comprehensive their data entry became. In this case, friction occurred between a person and a tool.

One possible area of friction for Protectus that we didn't mention was the sales rep's interaction with a financial planner (friction between workers). How long does it take a financial planner to work with data and return proposal information to the sales rep? Do these types of projects interrupt more lucrative or more important tasks, annoying the financial planner? These are areas across the enterprise where friction occurs.

High costs

Ever come across a simple process with suspect costs attached to it? Costs don't always track with an associated activity. Workflows with high costs attached to them are usually hidden opportunities to cut spending.

Time and time again, we've gone over the high costs associated with printing materials. We've also touched on travel expenses. What about the cost of supplying your workforce with laptops? Mobility can accomplish the same goals (often even more efficiently) using a $200 phone over a $1,200 computer.

Wherever costs pile up, take a second to analyze this part of the process. Ask yourself: "Can I make this workflow better with mobility?"

Repetition

Repetitive processes have been a part of business forever. Repetition can be good—it's how you learn to be great at what you do.

Repetition can also be bad, especially when you end up doing the *same* work twice. One of the most glaring examples of bad repetition is recording information on that notepad or clipboard and reentering it later on in the day, week, or month.

In our grain silo example, we saw a relay of information. This work was repetitive. It included a middleman, who personifies repetition. Anywhere you find a middleman, you're likely to find extra work that doesn't add value. (You can even think of your notepad or clipboard as a "middleman.")

Interruption

Who doesn't hate interruptions? They ruin your train of thought, take you and anyone you're interacting with out of the moment, and severely hurt your chances of finishing anything productive in a reasonable amount of time.

Unfortunately, interruptions are common across the enterprise. Protectus saw interruptions to their interaction with a potential customer, hurting their conversion percentages. Almost all customer interactions could benefit from fewer interruptions, in fact.

Interruptions can also occur when workers have to wait on other workers for their piece of a project. It's impossible to eliminate many of these instances—but it is possible to make the process simpler, helping you solve the problem of...

Long waits

These days, we can barely wait for a computer to boot up. Our patience with other devices is equally thin.

True, the boot-up process of a computer is far from meaningless. A Protectus sales rep waiting on a financial planner's expertise isn't meaningless either. But the process could still be accomplished more quickly, making the workflow more effective too. Customer interactions may be the worst places for long wait times. Therefore workflow revision can become a huge factor for customer service and support. Take any retail interaction as an example: do you want customers facing long lines at the register when you could have a service representative walking the floor, checking them out on the spot?

Frequent mistakes

Everyone makes mistakes. It's part of how we learn as humans. Mistakes help us grow professionally. Ask entrepreneurs about mistakes, and they'll list off dozens of their biggest.

The point of mistakes is that we should learn from them. If the members of your workforce continually make the same mistake twice, chances are good that a part of the process is setting them up for failure. These mistakes also indicate a workflow that requires more intelligent tools.

Frequent mistakes in data entry are an obvious example. Confusing software platforms and repetitive processes often cause these mistakes. Find out what's causing frequent mistakes in your enterprise, and you'll have a great opportunity to create more efficient workflows.

Too many cooks

How many people are involved in a given process? More often than not, the answer is "too many."

We get so used to certain processes that it's tough to question whether you even need another worker involved. This is certainly the case with Protectus' use of financial planners. It was also the situation for the grain silo trucks.

Are too many cooks spoiling your pot? Wherever more than one person is involved, ask yourself if there is a way to simplify and let an app support the person instead of pulling in additional resources.

Coupled with the right perspective, these strategies can help you understand where workflows go wrong, offering you a strong basis for creating your app concepts. While you're at it, it's a good idea to take a hard look at the tools you're using to accomplish these workflows.

TAKING A LOOK UNDER THE HOOD

The machinery of your enterprise is complex and expensive. Sometimes it's irreplaceable. Other times, it's completely obsolete.

The hardware used in a workflow often indicates workflow issues that require evaluation. Mobility can be the improvement or the replacement to the existing hardware. In many cases, hardware consists of single-purpose devices.

In the last chapter, I suggested using tools hands-on to get closer to workflows. Here, let's take a look at what makes tools practical or impractical. Andy Graham's story can help us understand how tools and devices evolve.

Andy built his livelihood introducing new mobile-friendly hardware into hundreds of enterprises. Since 1999, his company, Infinite Peripherals, has helped organizations equip devices with mobile printers, bar code scanners, signature capturers, mag-stripe readers, and more. "Organizations are making their workers more productive with devices that pull up real-time data," Andy says. "We create complementary hardware to help fill the limitations of those devices."

Andy is close to the enterprise mobile experience. Over the last fifteen years, he's watched dozens of organizations make the leap to mobility.

One major retailer that Infinite Peripherals worked with started using a mobile app as an inventory tool. Requests for other apps suddenly began to flood in—for tracking time and attendance, scheduling, planogram generation, geographical product lookup, customer loyalty, and more.

According to Andy, this functionality is at the heart of hardware replacement. "You have existing, single-purpose devices out in the field. Sure, they may be durable. They probably accomplish the one thing they were built for. But what else can they do? That's where mobility comes in handy." Today, technology convergence is the norm for consumers. Why would your workforce expect anything less than a multifunctional device?

Andy says retail has been one of the biggest success stories here over the past two or three years. He also sees it happening in warehousing. Healthcare has a huge opportunity to make electronic medical records and regulation compliance mobile. Even nonprofits are making waves in mobility.

Mobility owes much of its success in these fields to its ability to replace one dimensional hardware. View your mobility project as an opportunity to reevaluate the hardware that powers your enterprise.

Here are ten questions you should ask as you go about evaluating the hardware involved in your enterprise processes.

1. Is it even necessary?

Even the obvious questions need answers.

2. Is this hardware obsolete?

Your industry is moving forward. That means some hardware gets left behind. Why install GPS tracking devices on truck fleets when you can manage logistics in real time through a mobile device? Why buy key fobs when you can configure smartphones to unlock campus doors?

Obsolescence is a clear first consideration when you evaluate your hardware. An organization with foresight doesn't wait until hardware is rusting—it acts before it meets these problems. (Urban Outfitters, for example, is one of the retailers pushing forward on eliminating point-of-sale checkout lines through mobile payment processing.)

3. How much does it cost to maintain?

Over time, enterprise hardware naturally undergoes wear and tear. Some of that hardware requires maintenance. And maintenance isn't free.

Expensive devices and machinery can cost hundreds of thousands of dollars a year just to keep them running smoothly. It's worth keeping an open mind about alternatives where possible.

4. How easy is it to use?

People just don't want to use hardware that doesn't fit their expectations. When your expectations are shaped by smartphones, it makes a lot of enterprise devices seem hard to use by comparison.

A cryptic, confusing user interface is the last thing you want to put in the hands of your workers. Yet we still use legacy devices that are decades behind smartphones in terms of interactivity. It's time to start thinking simpler.

5. Is it heavy, cumbersome, tough to lug around?

Remember those pesky luggables? Plenty of organizations are still forcing workers to haul hefty equipment from place to place. Some jobs still require large devices. (A riding mower, for example, is today's most effective device for cutting a large lawn.) For those that don't, there's an opportunity to make a major, cost-cutting change.

6. Is video playback or image viewing its purpose?

It's time to throw out those old televisions and start beaming video back to your mobile device. In many cases, devices with screens that can't fit in your pocket have become obsolete.

7. How many things does it do?

A device that does only one thing is a thing of the past. It's time to move beyond single-purpose devices and into the functionality convergence of the future.

Even if it is multifunctional, what percentage of the device's functions do workers actually use? That number might be surprising.

Not every piece of machinery is multifunctional. If it has computing power, though, there's a good chance it could open a world of practicality.

8. How long does it take to train people on the device?

Training can be a huge investment, especially for old, clunky, confusing devices. Why spend time and money training when you can hand someone a device that can be up and running in seconds?

9. Can a mobile device replace it?

Those of us who aren't as close to mobility can only imagine a handful of practical ways it can help us. Looking at the challenges first helps us open our imaginations.

With the issues in perspective, the usefulness of a mobile device can be seen in a whole new context. A little creative thinking can lead to great ideas on how a mobile device can replace existing hardware.

10. Can a mobile device control it remotely?

Even if a mobile device can't replace hardware, it can become an indispensable complement. Consider again the truck driver controlling the flow of grain from the silo with a phone.

Such processes are not the only places where remote control comes in handy. Building managers can control lights, climate, and other factors directly from their phones. Restaurants can warm up ovens. Factory workers can manage mechanical assembly lines. Stage managers can control lights and sound. If you can attach a Wi-Fi or Bluetooth receptor, you can control it from your mobile device.

Think of your potential mobile devices as a source of renewed hope in your organization's hardware. Spend time taking stock of your hardware and finding its limitations and holes that mobility has the potential to plug.

Now that we've taken an overview of the situation and the problems, it's time to move forward on creating the right solution.

CHAPTER 9
Step Four: Generating App Ideas

Step 4 of the BDA process is all about sitting down at a conference table next to a big whiteboard and shaking a bunch of app ideas out of your system. The flow of concepts may start as a trickle—but before you know it, they'll be pouring out of you like a waterfall. The more concepts you can amass, the better.

Use the workflow challenges that you dug up in Step 3 to drive app concept development. The list can include flashy ideas you got from other organizations too. But in the long run, you'll want to choose the concepts that expressly meet your needs.

It's very easy to take a look at the Protectus example and see how it could apply directly to your organization. But Protectus is a fictional company, and while a Protectus look-alike is sure to exist, you shouldn't jump to that conclusion until you've completed the steps of the BDA. That way, you'll have the proof you need to build a business case for yourself, your committee, and your organization.

What exactly is mobility capable of accomplishing within your enterprise? The possibilities are abundant. It's up to you to come up with the unique mobile value propositions for your audience.

MANAGING THE BRAINSTORMING PROCESS

We've all spent some time in the war room. The ideation process should be a structured couple of days of informal meetings.

As the change agent, you'll probably have a lot of legwork done beforehand. This is your opportunity to catch everyone up, get some of the core concepts fresh in their minds, and start pumping out organization-changing ideas.

For simplicity's sake, we chose eleven examples of app concepts identified by Sam and Protectus. The truth is that you'll want to get a whole lot more out of your ideation sessions—roughly thirty concepts—by the time the ideation process is over. That way, you have a good pool of ideas to test against each other.

Below are five steps to help you create the right framework for brainstorming.

1. Appoint a moderator and scribe.

One of the most important elements of your committee is a strong moderator, someone with great leadership skills who gets mobility. This may be you, the change agent. It could be an IT director who is in the know. It might even be the project manager on the vendor side. Whoever it is, someone who understands the BDA process must take charge of these sessions.

The scribe is responsible for capturing the concepts and keeping the content organized. A couple of large whiteboards go a long way for collecting the ideas.

2. Share the prep materials and refresh everyone on the goals, workflows, and issues.

Before getting everyone together for the brainstorming session, make sure to hand out a summary of the goals, workflows, and issues that have been captured so far. Once the team has had a chance to familiarize themselves with the material, bring them together and get started by walking through the organizational goals that have been laid out for the year. Those unfamiliar with the goals need to hear them. It'll be a good refresher for team members who are already familiar with them, and they'll be points of reference later.

Next, walk through the workflows that support those goals and the issues identified with each. Take any issues that jump out at the group as important and get those up on the whiteboard so they stay visible during the brainstorming process.

3. Brainstorm away.

It's time to get down to brass tacks. Continue the discussion with a running list of the app ideas that come up. Don't discount anything at this stage. The key is to get them all down; you can trim the list later.

Here's a brief rundown on how to go about eliciting app concept ideas in the brainstorming:

- Look at each user role involved in the workflows and see what challenges each has that apps could address.
- Walk through each issue and ask yourself what an app would need to do to solve one or more of the identified issues. In other words, what would the purpose of the app

be? What capabilities or functionality would an app need to address the issue?

- What information could an app give a user to address a workflow issue?

Remember to keep ideas lightweight. And, each app concept should represent a single function, not a suite. You'll create suites later on.

4. Fill in the Details

Your list of app concepts should be built out to at least twenty or thirty by now. Now add details to each app concept.

Who does it help? What challenge does it solve? Who interacts with it? What software and hardware platforms would the app need to function? Find the angles on each app concept and capture details. After the brainstorming meeting, distribute the app list and descriptions for group review. It's very likely that team members thought an app name meant two very different things.

Miscommunication on concepts can be tough to set straight further along in the process. To prepare for that possibility, appoint a champion from the group to take the rough app concepts and define them further. What does it let the user do? What issue is being addressed? What won't it do? Write at least two or three sentences about the app so everyone has a common view of it for later discussion.

You should also start seeing the overlaps—where certain app concepts are dependent on others. A price-quote tool, for example, might require a research app that pulls the information necessary to get the quote. Map out these dependencies so you can have a full picture of how different apps will work together.

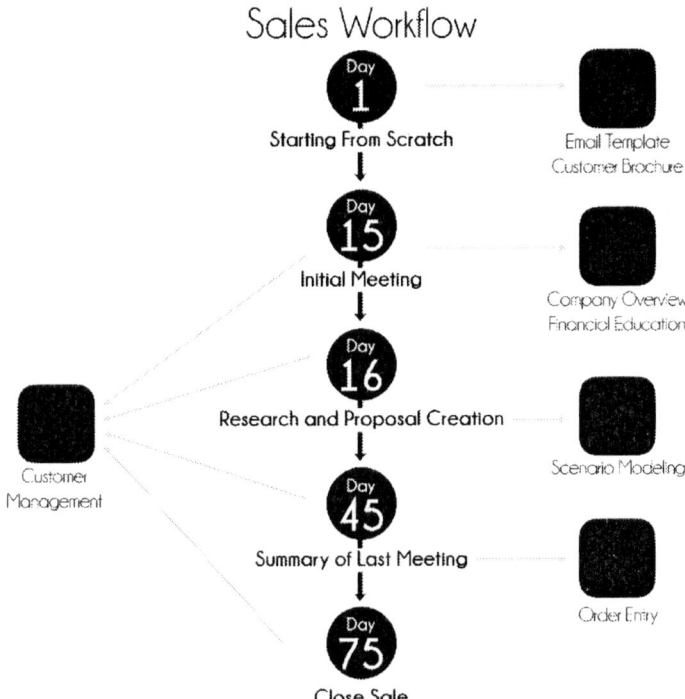

5. Group complementary apps.

This part of the process is meant to build out potential suites. While each concept could end up in more than one suite, it helps to categorize (or "tag") apps as concepts that have logical relationships. Make sets of loose categories to create suite concepts.

This is an exercise in understanding how certain apps could work together; one might function very differently without the other. The end result of this step is a list of apps and app suites you can now analyze.

CHAPTER 10
Step Five: Calculating the Impact of Your Concepts

Through the story of Protectus, we demonstrated how change agents calculate an app concept's potential impact on the bottom line. This is an incredibly important part of building a business case to secure budget and get executives and managers on board with mobility.

In Step 5, it's all about calculating the impact of your app concepts. Note though that finding low-impact apps is just as important to the process as identifying the high-impact apps. Just because an app concept won't have a huge or immediate impact doesn't mean it should be scrapped. Rather, it should be moved down the list to be dealt with after you develop the most significant opportunities.

We've already covered the equations you need for this step. Here, we'll take a closer look at the methodology behind the equations to give you a better grasp of why these numbers are critical to app development, and therefore an app's potential impact on your organization.

WHERE DO THE NUMBERS COME FROM?

For each equation in Step 5 of the BDA process, you'll need to take into account everything that could potentially go into the equation to get the most accurate result.

To pull the right number, it helps to know exactly where it comes from. Here's a breakdown of each of the numbers involved in your app concept evaluations.

NEW REVENUE POTENTIAL (NRP)

The NRP measures the potential for an app to generate new revenue for your organization. You can trace much of the new revenue back to the department for which you're developing your app concepts.

As shown in our Protectus example, here's how we represent the New Revenue Potential:

$$I \times T \times R / (100\% - I) = NRP$$

For example, assume a sales team spends 50 percent of its time closing deals that total $1B. When assisted by an app that can improve the efficiency of that process by 10 percent, their resulting NRP is $55M of new revenue.

Further, the (100% - I) part of the NRP equation drives the scalability of a workflow when the efficiency is introduced. If we cut the time in a workflow by 10 percent, it gives our team the time to create 11 percent more than its prior output. If we cut the time in half, we can hit 200 percent the output. If we reduce the time by two-thirds, we get three times the output. And so on.

The (100% - I) portion also adjusts the result so that we are looking only at the " n e w " part of the revenue. If we're going to add 50 percent revenue, the NRP should be 50 percent, not 150. We add the dividend to the equation to get a nice, solid percentage for additional revenue potential.

The numbers here reflect how making a given workflow more efficient frees up time for activities that can actually help generate revenue. Unlike the CSP, it shows what you can do with a sudden burst in productivity.

COST SAVINGS POTENTIAL (CSP)

The CSP helps you see what costs an app concept could cut without considering the potential for new revenue generation.

Put bluntly, pure CSP represents how much money your organization might save cutting the hours your workforce puts into their jobs. This number doesn't necessarily consider what happens when you allocate those workers or those hours to other tasks. (That's where the NRP comes in handy.)

The CSP also represents the time and money an organization can save by eliminating costs like paper and ink, travel expenses, shipping, hardware repairs, inside and outside consultants, administrative work, and anything else an app concept might reduce by simplifying the current workflow. You factor the costs (represented by C) into the equation, multiplying C by the impact of the app and the time spent on the process in question:

$$I \times T \times C = CSP$$

Note that the "T" in CSP as expressed as a percent of a worker's total time. The cost, "C," indicates the salary your workforce draws as it engages in the process plus any of the other costs mentioned above. You should look at cutting costs here in terms of being able to apply existing labor to other valuable tasks rather than seeing it as about downsizing. It's about making workers more efficient so they can do a whole lot more for your organization.

For our example app's CSP, we'd be looking at sales labor costs of $50M. At 50 percent of work time on this example process and a 10 percent impact, we get a potential cost savings of $2.5M.

Along with the NRP, the CSP is a factor that goes into generating the ROA. For Protectus' CSP, the hypothetical use case dealt with the sales department at an insurance company, the organization's primary revenue generator. However, defining your operations teams may not be so cut and dried. That's why you must always consider the next factor, OI.

ORGANIZATIONAL IMPACT (OI)

You might consider the OI your free radical, your floating variable. This could include items like eliminating a recurring cost (paper and ink, for example), benefits that go beyond general efficiency and time saved (like Protectus' eliminated meeting that should result in a higher close rate), and other intangibles (surpassing a competitor or boosting your stock, for example).

In the Protectus example, we saw how skipping a step meant more than just saving the time it took to perform that step. It added an entirely new measure of effectiveness, a factor that could have a major impact on the bottom line by keeping the attention of a chunk of the original lead pool.

This is what happens in situations where you significantly cut trips between destinations. It either means you can do remotely what used to require an in-person visit, or you can save steps that once stretched a process out on a longer timeline. In both situations, you have a process shift that can make a huge impact on the success of your workflows. That impact becomes a part of the OI.

OI can also include the "wow factor" impact. The wow factor of apps is often the reason organizations approach us about enterprise mobility in the first place. They see mobility as a shiny new toy. Taking the underlying story and BDA groundwork into account completes the package.

What can the wow factor do for your organization? Can it reengage a drifting internal audience and reduce employee turnover? Will it become a major differentiator to help you with recruiting top talent? Do your best to quantify your desired wow factor to get an accurate picture of your OI.

This part of the equation can seem less scientific, but it's still rooted in finding the right numbers. You can then couple the OI with the ROA to get an accurate perspective on the overall impact of your concept.

RETURN ON APP (ROA)

As we saw in the Protectus example, the ROA represents an accurate idea of what your company stands to gain from app concept development.

The ROA is a trade-off between using fewer resources to get the *same* output and keeping resources the same to generate *more* output.

As such, the equation looks like this:

$$ROA_1 = OI + NRP$$
$$ROA_2 = OI + CSP$$

Your ROA should fall between ROA_1 and ROA_2. Unless your numbers are wildly off, the ROA should give you an accurate picture of what the organization stands to gain in tangible assets.

It isn't all about the ROA, though. There are all sorts of qualified benefits and factors that go into making the right decision on your app concepts. This is where the final step of the BDA process helps you make a decision that weighs everything important to your process.

CHAPTER 11
Step Six: Scoring and Prioritizing Your Concepts

The final step of the BDA process is to prioritize the concepts based on criteria your organization considers important. It examines factors like NRP, CSP, OI, opinions of the end user, viability of development, budgetary restrictions, time to deployment, existing systems, and other factors to help you attack your mobility project from every angle.

Answers for certain criteria won't always be measurable. It's your job in these cases to make qualitative comparisons using the scoring methodology on the 1-to-5 scale as shown in chapter 3.

Your endgame here is a total App Success Score for each app. You find it by adding each criteria rating for the app. Sort the final list in descending order. The top third of the apps on your list should be your targets for the first phase of enterprise mobility.

Based on twelve different criteria, our client A.M. Castle and Co. was able to identify four apps with the highest scores. Together, these apps targeted between $223M and $323M in ROA.

They included:

- A customer-facing order status lookup tool that empowered customers and cut down on a major inside sales time investment
- A proactive alert tool that helped guide the customer through an error-free ordering process
- A test report search tool for inside sales reps that cut their time spent on research
- A quote creation tool for inside sales reps to pull pricing numbers much more quickly

The questions Protectus covered are pretty standard scoring criteria that we use with our clients. This step is totally customizable based on the requirements of your organization, the restrictions of the project, and the limits of your imagination.

As a refresher, here are the criteria Protectus evaluated in Step 6 of its BDA process:

Time to realize ROA

Most app concepts won't realize their ROAs in the first few months. Some won't get there in the first year. Concepts that move more quickly obviously get a higher ranking here. Some apps have nearly immediate returns.

Cost and time commitment

Some apps can be cost-restrictive. Others fit that sweet spot in your budget. Take into account development, hardware, training requirements, etc. App development can take some time, especially if it requires integration with backend systems. This can impact an app's score. The less time and money you have to invest, the higher the score.

Your IT team or development partner can help put together quick, high-level estimates. You don't need specific development budgets; you just need a way to compare which apps will be more complicated to create than others.

Amount of training

If an app requires extensive retraining, the concept may not be intuitive enough for consideration. App concepts that require a smaller training investment get higher scores.

Number of workflows addressed

An app that addresses one issue is often less valuable than an app that addresses two. More areas where an app can have an impact mean more efficiency and a higher app score.

Risk

What kinds of risks are associated with app development? Is the concept a heavy investment for a potentially low return? Is your oversight team less enthusiastic about this concept? Weigh risks to find an accurate numerical rating. The app concepts with the lowest risk earn a score of 5.

Mobile API

Connecting mobile apps to enterprise technology systems can be easy or hard depending on how open or accessible the business systems are. Do they have documented application programming interfaces (APIs) that allow remote software to link to them? If you are hearing buzzwords like "RESTful web services," "service-oriented architecture" (SOA), "Extensible Markup Language" (XML), or "JavaScript Object Notation" (JSON), those are good signs that the business systems have integration points.

End user excitement

This is where your workforce's feedback comes into play. If workers want the app or app suite, you'll obviously have a much easier time working it into the equation. Exciting apps get higher app scores.

THE RATING PROCESS

Have each of your participating team members rate each app for every one of the agreed-upon criteria on that 1-5 scale. People should feel comfortable with what qualities the scale rates. As long as each person is consistent across the apps they rate, specific details won't matter.

    The scoring can be done on paper and collected by the scribe for collation, or a shared online spreadsheet can be used. I've had good success using a simple, online survey tool like SurveyMonkey (surveymonkey.com) to collect information from people between sessions. That way, they don't have to be present to wrap things up.

We saw how at Protectus, Sam ranked each of the apps on the desired criteria and added the columns for each concept, finding their app scores, ROAs, and OIs to help guide her decision on choosing the right apps.

Other criteria you may want to look at instead (or in addition) are:

- What kind of time commitment does IT need to put into this app?
- How easy will it be to get executive buy-in?
- Is this app a competitive differentiator?
- Will implementing this app be a good template for future mobility?
- Will ongoing costs to keep the app moving forward be low?
- Can this same concept be applied inexpensively to other lines of business?

Asking the right questions will help you get a better handle on your app concepts. Again, just because a concept doesn't get a top score doesn't mean you have to scratch the idea; the scorecard is meant to be a roadmap of sorts—a prioritized list of app development projects.

Take the top third of your resulting list to find your phase-one cutoff. The next third is phase two. Anything lower than that falls in a nebulous "later" bucket. Recheck the list for sanity and common sense, though: if an app that looks like it should be done earlier is near one of the phase boundaries, move it up.

Another area to look at is how the breakout for the app suites and dependencies line up. If there are a number of phase-one apps that require a phase-two app to really deliver value, that phase-two app has to move up as well. It may also make sense to group apps in a suite to deliver a completed workflow rather than wait for some of the components.

When you hear someone contribute something like, "But my app didn't make the cut," make it clear that the list isn't arbitrary. Explain that the goal of this process is for the most meaningful and impactful apps to climb the list and for obvious, less impactful apps to fall further down. Be wary of arguments like "We already promised app-that-is-way-down-the-list to someone." It's not uncommon for an executive to contribute an obvious app idea for his department without really understanding what the impact (or lack of impact) would look like.

Using some very rough estimates of the scope of the apps, phase one and phase two can be mapped out on a calendar. These are typically six-to-twelve-month cycles, each containing five to ten apps.

With this roadmap, you'll make your concepts into reality. To get there, you'll need some guidance on how to manage the project.

CHAPTER 12
Considerations for App Development and Launch

You've built your roadmap, chosen your apps, and convinced your executive team that you're doing the right thing. Now it's time to make things happen.

App development is a skillset all its own and requires a development team with extensive experience building that skillset. A lot of moving pieces go into building your Billion Dollar Apps. It's OK (and often in your best interests) to seek outside help.

As the change agent, you may be tasked with managing the development process. Or you may be part of the advisory team working with the developers. Whatever your role, it helps to understand how the process unfolds. That way, you can predict obstacles, plan for limitations, and prepare for emergencies.

Consider a few hypothetical scenarios (some more likely than others) where the app development process could go very wrong.

Your development partner can't meet expectations.

This could happen for a few reasons. In many cases, your advisory team may not understand that certain expectations are unrealistic. Others may have engaged the wrong development firm—one that is fraught with inexperience, lack of leadership, or too many clients.

Your mobile team is moving faster than the rest of your team.

Who should be driving the development process? What happens when one team moves faster than another? There are ways to stop this, but it can be more effective to plan for this possibility, making sure you're prepared for it.

Your project team and stakeholders are out of sync.

Without a shared set of expectations, teams involved in different elements of the development process can have very different outlooks. How do you ensure people are communicating? What happens if business goals aren't aligning with project outcomes?

Without solid oversight, app development can go wrong in dozens of places. I'll teach you how to avoid these mistakes by creating the right framework for success.

It consists of a six-phase approach:

1. Plan
2. Define
3. Design

4. Build
5. Distribute
6. Support

First, we'll look at things you need to know to create a winning strategy for app development.

PHASE 1. PLAN

After your mobility project gets the green light, a well-managed development process makes all the difference. From choosing a platform to engaging a development team, building your Billion Dollar Apps requires a careful but deliberate hand.

Here we'll take a look at six different aspects of planning the development process.

1. Communicating guidelines
2. Forming your development team
3. Creating a developmental roadmap
4. Establishing device policies
5. Defining security & access control
6. Preparing for distribution

First, it's essential to write down the top-level guidelines for project management.

COMMUNICATING GUIDELINES

Why is this phase of mobility development important to the success of your company? You've spent a lot of time building a business case for your app(s). But have you summed it up in a way that will help unite the team that works to make each app concept a reality?

Create a document containing expectations and guidelines. These will act as the glue that keeps your advisory and development teams motivated toward the same ends, ensuring a better basis for ongoing communication.

Elements you'll include are:

Business case expectations

What's the motivation behind mobile development in the first place? What are the goals the company hopes to achieve? These expectations should be communicated to the entire team early on in the process.

Roles and responsibilities

Clearly define the purpose of each team member (before the project gets started). It'll save you a lot of headache down the road on accountability.

Creative brief

A short description of the project, along with end-user profiles, can be a helpful piece of content to keep the project on track.

Deliverables

Briefly explain each product of the project and how to accomplish it. Now that you have guidelines, you'll need a team that they'll govern.

FORMING YOUR DEVELOPMENT TEAM

Your development team consists of two major branches: an internal advisory team and an internal or external development partner. These groups work hand in hand as a coordinated development team all the way through your mobility project.

Your advisory team should consist of a handful of stakeholders in the project. Many of you have experienced the confusion of working with too many people; it's best to keep your advisory team small—but it should include people who fit some very specific requirements. In addition, understand that it helps to have a few different points of perspective among your team of internal advisors as you go about the development process.

I won't limit each of my recommendations to actual titles. Instead, the list here reflects the perspectives that are essential to a well-organized team of advisors. This team focuses on project oversight and on collaboration with the development partner.

A change agent

By this point, you know something about the role of the change agent. This is your role. You'll bring the ability to tie together clout, perspective, and knowledge of the BDA process.

A high-level manager

Who fills this role depends on where the change agent lands in the enterprise. An IT change agent requires a manager specific to the workflow the suite of apps serves to simplify. A change agent in another department typically needs an IT director, CIO, or other manager in the IT department.

A direct manager of end users

Whomever the end user reports to directly is the right person to fill this role. (In Oldcastle's case, this would be a foreman. Protectus might require a local sales manager.)

An end user

This one is a bit tricky. You don't necessarily want a single end user sitting in on every strategy session—but you do need him or her to work with early versions of your app suite. This role may also require more than one person to fulfill its demands—you may roll out new apps to a test group, for example.

A marketing or communications resource

Someone with marketing and communications skills plays a necessary role in ensuring app adoption. This person is here to help drive the value of apps through internal or external campaigns.

With your advisory team up to speed, it's time to start assembling your development team. This can be a much trickier process. Most organizations will want to start in-house. If you can't find the right skillsets internally, you may need to extend your search to outside development partners.

The right development team includes a very deep skillset, whether it's an internal or external team. Mobile development is much more involved than just building an app that displays on the screen. It needs to talk to enterprise systems and function in the real world, where connectivity can be erratic. The skills include everything from understanding the user experience to deep integration with systems.

As you search internally for the right team members, you should look for seven critical characteristics. As I noted, if you can't find these in-house, you'll need an outside development partner with this mix of credentials.

Process analysis

Your developers should have experience digging into and understanding your business's workflows, regardless of their experience in your industry.

User experience strategy

Expertise in UX is crucial for defining the simplest paths the end user can take through the app. It's the methodology and architecture behind making processes more intuitive through technology.

User interaction design and build

Great app developers understand how design affects how users interact with their apps. They can design and build a strong platform that leads the user through the desired tasks.

Core and server logic

These two disciplines define how apps talk to a backend system and vice versa. A strong communication backbone is necessary for a reliable suite of apps.

App communications

This discipline refers to how an app interacts with data outside of the mobile device. Put simply, it is about how information is broken down for more efficient app communication.

Web services

"Chatty" web apps eat bandwidth and slow down app process. Web services specialists streamline data communication, helping create an optimal user experience.

Data

Understanding how systems play together and how data flows through each of them is usually necessary for the right combination of apps.

An app development team with deep experience may even create its own middleware—a software bridge, of sorts. This type of middleware simplifies how an app pulls data from the backend platform. It can be a huge time and money saver throughout the development process.

If you do end up evaluating third-party developers, you'll need a strong set of additional criteria. Who you choose to work with may be one of the most important decisions you'll make during the development process.

The right partner:

- Has extensive experience in app development
- Understands how to organize workflows into great user experiences
- Can offer glowing recommendations from previous clients

- Aligns its brand and goals well with yours
- Understands and practices the tenets of the BDA process
- Travels across the country to meet face-to-face
- Is willing to tell you, the client, when you're wrong, but—
- Compromises when it makes sense

The wrong partner:

- Just says "yes"
- Makes little effort to build a long-term rapport with you
- Tells you something is impossible (rather than explaining what it would take)
- Consistently misses deadlines
- Is difficult to get in touch with
- Takes no time to explain strategic development decisions
- Revises pricing or timing mid-project with no changes made to the project scope
- Meets you once and only works remotely following the initial meeting

You should always meet with a potential development team before you form a relationship. If its people are not willing to go the extra mile, they're not worth your time.

You've built out a strong team for the job. Next, it's time to plan out the development process.

CREATING A DEVELOPMENTAL ROADMAP

With any major project, it helps to form a plan before you start building. Your mobility project is no different. A developmental roadmap helps govern the app construction process.

With everything on paper, you can establish trust and ensure a great working relationship. You'll also have a clear path by which to define success.

Here are four tactical elements you should plan to work into your roadmap:

Timing

What are the major milestones? When are deliverables expected? How long will the project last? (Hint: A BDA development project should spend four to twelve weeks in development. Anything longer may be a clue that you're trying to do too much.)

Goals

Setting goals along the project roadmap ensures development stays on track and helps measure success.

Budget

Don't get caught spending more than you bargained for. Be sure to define the full scope of the project's budget prior to jumping in.

Communication guidelines

Defined points of contact, a main communication medium, scheduled in-person meetings—these are all elements of a successful project and should be set forth in communication guidelines.

With the scope of your initial projects framed, your next move is to create policies that govern device ownership and usage.

PREPARING FOR DISTRIBUTION

App distribution occurs at a few different phases of mobility, from beta testing to comprehensive enterprise-wide rollout. It's a good idea to look at your distribution options early so you can work this into the schedule.

You'll want to look at three separate options for distribution to internal and external users.

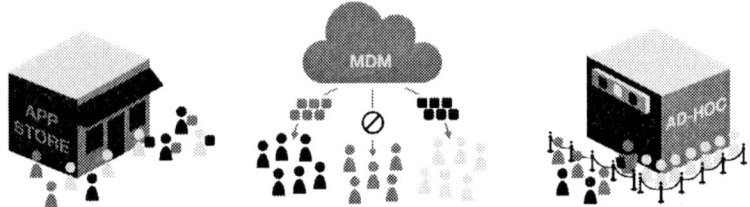

App Store
With app store distribution, you don't need to manage employee devices directly. Instead, users have the ability to grab whatever apps they need on whatever devices they're using. App stores are good for dual app distribution, where access to certain features and data is limited based on user roles.

Mobile Device Management

Whereas app stores are open for external users, mobile device management (MDM) distribution only serves internal devices preloaded with the software and therefore gives you a higher level of administrative control. With MDM software, you have the ability to push apps out to people and check for version compliance.

Ad Hoc

This type of distribution is most often used for testing, distributing to developers and alpha/beta testers. You might use this method to distribute apps to a limited subset of users for feedback. For ad hoc distribution, organizations usually work through third-party services that make it easy to put apps in the hands of the right people, especially users outside the company.

With everything in place, it's time to define the requirements of your audience and your apps.

PHASE 2: DEFINE

A lot goes into the development of an app after you have your concept. Before you design and build your app, you should walk your teams through certain expectations. The audience and the technology define them. Your users drive some expectations; others are just a part of the systems already in place. Wherever they stem from, they must be defined prior to jumping into app development. Developing app expectations is a two-step process.

CREATE USER PROFILES: PERSONAS AND STORIES

At this point in the game, you've gathered tons of insights on your prospective end users. Now it's time to build this into functional profiles of the people for whom you plan to develop.

A user profile consists of:

A user persona

Understanding app expectations starts with building an accurate picture of your users. What do you know about their ages, levels of experience, special needs, and so on? Build accurate personas, including details about the person that are relevant to his or her job.

A user story

The user story is a one-sentence description of the action a set of users takes to reach a certain outcome. It's usually represented as "User A does X so that he or she can accomplish Y." For instance: *A workman fills out a time sheet so he can get paid.*

You may need to repeat these for different roles included in the app concepts you're developing. Creating these segmented user roles simplifies the task of role-based distribution when you're ready to deliver apps to your workforce.

PREPARE FOR SCALE AND PERFORMANCE

To prepare for app development, you need more than just qualitative information about your user. It helps to understand scale and performance considerations you're already working with in terms of existing technology.

To do this:

Look at scale requirements.

How many people will use a given app concept? How much data will flow between the app and the backend system? Writing down details on the scale of your apps' reach will help you plan for the development process.

Understand backend systems.

What backend systems will your app need to talk with? What technical limitations do those systems bring to the table? (A common issue you might run into is a poorly defined API on the backend.) When something goes wrong, how should the app respond? What values do you assign in specific fields? What parts of the backend system's connections should your apps use, and why?

With a well-rounded understanding of audience and technical needs, the next step of the process is to design a truly compelling user interface for your apps.

PHASE 3: DESIGN

We've made no secret about the power source behind the world's most successful Billion Dollar Apps. If you still feel like you're in the dark, I'll take a second to reiterate it here: **Billion Dollar Apps are all about making life easier for your end user.**

User experience (or UX, as it's known to most designers and developers) is a term that encompasses the journey of an app's user. It's about putting your ego to the side and realizing that your app was never just about what *you* want. Developing apps that effect real change in your organization is always, without fail, about doing what *they* want.

A few years back, software developers would add features as a way to move their platforms forward. Along the road, someone realized that ease of use was more important than features when it came to creating solutions that his or her customers loved. The discipline of UX was born. UX now plays a huge role in how developers approach the creation of all kinds of software and hardware. Great websites are built on a solid UX foundation.

The user experience includes all of what people who use your app will feel and do when they dig in. (Don't confuse UX with user interface [UI]. Your app's UI consists of the device and the software that power the UX.)

The right development partner puts UX at the core of its skillset. As part of the project team, you should know a little about how user experience comes into play too. UX choices don't always make the most sense visually, but the right methodology behind it is critical for ensuring the end user leverages the app to its potential.

Here are eight things you should know about UX for app development before you start your BDA project:

1. UX has a lot to do with the path an app user navigates from start to finish.

The user experience is heavily defined by how your user gets from point A to point B. UX experts spend a lot of time understanding user motivations, building paths, testing apps, revising paths, and retesting apps.

When your user asks, "What next?" your app should have an immediate answer. Otherwise, you risk losing the user's attention. Without it, your app can't become a powerhouse force within your organization.

2. UX also has a lot to do with touch.

Ever tried to click a link on your mobile device, accidentally clicking a different link in the process? Websites and apps that aren't optimized for mobile miss out on something crucial: they don't take "touch" into account.

Mobile apps are designed with "fat fingers" in mind. UX strategists understand that it's more important to have accessible buttons than it is to cram a bunch of different pathways into a mobile app. This approach takes a lot of frustration out of the process, easing the app's user experience.

3. Web designers often don't have the right UX experience.

Unfortunately, you can't trust just anyone to create a phenomenal user experience. A compelling UX should be left to app design professionals.

Web developers tend to lack the experience and skills necessary for your mobile app because they're used to designing for the desktop experience. The web was fundamentally designed to function with a mouse, and that's how many designers approach it.

They haven't always been trained to focus on touch, information architecture, or UX. Many web designers focus on visual design over user flow. They just don't fit the bill when it comes to designing a compelling app experience.

4. Value and ease of use are direct goals of UX strategy.

It bears repeating that your app's end users only care about two things: the value the app offers them and how easy it is to obtain that value. UX strategists are concerned with both of these elements.

You already get a head start on providing value when you decide which app concepts make the most sense for your business. Your development partner will help you narrow down what's valuable even further through testing and quality assurance. That way, you can be sure your app delivers value that your end user can actually work with.

5. Apps that accomplish these goals result in strong engagement.

UX is so important to the process of app development because it focuses on how best to engage an app's audience. The perfect combination of value and simplicity results in apps with excellent engagement levels. This makes usage more effective and comprehensive, resulting in a higher app value for accomplishing organizational goals.

Engagement is critical for the success of your app. UX is focused on understanding engagement from the second users log in until the moment they exit the app suite.

6. More features result in weaker engagement.

The old adage "less is more" applies well here. App simplicity is better: fewer features, not more. The more functionality you add to an app suite, the better the chances you'll confuse your end user. As we well know, a confused end user is not engaged.

Trust your development partner here. Members of your team may want more from your app suite, but it's more important to focus on the most important elements first to get users familiar with your apps. Then you can add functionality where it's necessary.

7. Eye-catching design is important. But it must be simple.

Dull design robs your end users of some excitement, but flashy design can be dizzying. Eye-catching design falls somewhere in between, giving your app a sleek look without overshadowing its intended purpose.

UX strategists focused on the design side of things usually understand how design affects apps. Don't be misled by your internal team's desires to dress your app up or strip it down. Focus on a professional design that drives intended user behaviors.

8. UX encompasses everything, even less obvious app functionality.

The user experience extends beyond the realm of design and architecture and into distinct use cases. One great example of this is the importance of "ruggedizing" your app. A rugged app performs almost as well offline as it does online, storing information and housing enough data to help end users do their jobs.

User-experience design takes into account all of the elements of the end user's organizational role. It relies heavily both on the legwork you do before app development and testing throughout the development process.

OK, so your internal and development teams have partnered to envision a great app with an engaging user experience. But your job isn't done quite yet. Next, it's time to manage expectations for the app-building process.

PHASE 4: BUILD

Iterative (sometimes called "agile") development is a process that some of today's most successful software engineers embrace.

For our purposes, iterative development refers to releasing versions of your apps before they're finished to gather feedback from end users. That feedback helps shape the next version of the app, so that each is better than the one before it. Eventually, you end up with a highly effective app.

Your apps can spend their entire lives undergoing iterative development if you have the resources. That way, you can keep up with technology by evolving nearly in real time.

You should engage a development team that is comfortable with iterative development. This is the best way to get the most out of the app-building experience, ensuring you keep the user at the center of the process. Iterative development relies heavily on gathering data around how users interact with your apps. Testing is essential.

WHY QA AND TESTING ARE IMPORTANT

What would have happened if Google had gone to market with its search product and left it alone for years before updating features and launching a next version? In all likelihood, the company would not be the tech behemoth it is today, and you'd still be using Outlook to read your e-mail.

Quality assurance (QA) and testing are critical features of the development process. They help you standardize the pieces of planning and building that work. They improve the process itself (in addition to improving your apps).

All this can be quite challenging due to the diversity of devices and operating systems. Mobile testing tools aren't quite as sophisticated as desktop tools. Your team must define expected use cases and test those rather than trying to test everything.

WHAT IS APP TESTING?

App testing is the process of monitoring an app under controlled conditions, collecting the results, and analyzing them to understand how you can improve the technology. Testing occurs under both normal and abnormal conditions. The testing process usually starts with a hypothesis—that certain actions will find a certain result—and enlists real people to interface with the apps to see if the hypothesis holds water.

Any good product development includes testing, regardless of whether the product faces consumers or employees. Rigorous testing ensures that apps will hit their marks. It also helps you keep your apps up to speed when technological conditions have advanced.

Another reason testing is so important is that it helps plan for technological limitations that your app and devices may encounter. Understanding where your technology needs reinforcement stops these limitations from clogging up your support desk and (as a result) forcing another round of development after your devices and apps have been rolled out to your workforce.

By the way, QA and testing are also iterative. Your app team should continue to fix bugs as your apps evolve. See and touch your apps regularly so that issues aren't suddenly realized at the very end of the process, forcing you to backtrack and start over.

The development team you choose should be well versed in both QA and app testing. It should also become a cornerstone of how any internal IT or technical teams manage development projects in the future.

PREPARING YOUR BACKEND SYSTEMS

Backend systems usually aren't mobile-ready and typically take longer to modify than building the actual app. If you start building your app before you've had a chance to evaluate your backend system, you may be in for major headaches down the road. It's better to plan for backend integration before you jump into the development process.

Backend integration is one of the most common risk factors for enterprise mobile implementations. Don't make the mistake of being unprepared; check any backend systems that you want to work with your mobile app for mobile readiness.

It's very common for an app team to get stuck waiting for a new or updated web service. Throughout the process of development, it helps to sync the app and backend development cycles. That way, app dependencies are created before the app team needs them, ensuring the project follows its timeline.

There's a good chance that teams working on backend integration will take longer than those working on actual app development. This is a normal assumption because of the scale of backend systems and processes. One part of preparing for this lag is scheduling backend development early. Later, we'll talk about how middleware can help too.

Here are three tips you'll need for preparing your backend systems.

Stress Test for End-to-End Performance.

Some backend systems are unable to scale or don't scale well. The first tests for mobile readiness should check whether or not backend systems are scalable. Usually, this consists of end-to-end performance testing through stress tests.

One set of stress tests should focus on how many users the system can handle at once. Will too many people accessing the system slow it down or make it crash? Backend systems aren't always built to handle the load of people you expect to use them through mobility.

The other set of stress tests focuses on how much data can run through the pipeline. Multimedia and other large product data sets may take longer to transmit, clogging up the user experience with long waits and failed uploads and downloads. Stress testing data transfer helps you understand limitations and bring in reinforcements.

Build Sensors Into The Experience.

Sensors measure different variables within the app experience. If you build your app and configure your backend without sensors in mind, you'll have very little to help you gauge the success of your mobile plan.

Early on, it's important to take into account analytical tools that'll help you collect the data you need to understand app usage patterns. Sensors will become part of the frontend interface and the backend system, so it's important to keep them in mind as you make your backend mobile-ready.

You'll need sensors to understand how people use your apps, which segments of users are on board, who is lagging, how long transactions take, and where bottlenecks are occurring. Identify issues with app workflows based on how users are *actually* using your apps. Don't forget to measure the health of your app by collecting error logs and analyzing corresponding reports. With this information, you can prioritize updates and bug fixes.

Use Middleware.

Middleware is technology that forms a bridge between your app and your backend system. For app development, it gives you the opportunity to stub out backend system connections while they're being developed. When app development is completed, the middleware will link the app with the updated backend.

Middleware also helps reduce the burden on the app itself. As data flows back and forth between the app and the backend, middleware can trim the data down to ensure it doesn't overload the device or the connection with useless information. This results in a faster interaction and a speed boost to backend systems.

Finally, middleware plays a big role in the development process. If app teams are moving ahead of backend teams, middleware allows the app to function as it actually would in a live environment—talking to the software in the middle before data passes to the backend. This simplifies app testing even before the backend is ready.

PHASE 5: DISTRIBUTE

So far, we've seen a lot of reasons why mobility is one of the most intuitive technological developments yet to hit the enterprise.

We've also seen that Billion Dollar Apps are about far more than just mobile adoption. In essence, they're about changing the way we do business.

Change is scary. As a change agent, you know change can be a catalyst for better business, but it can also be difficult to embrace. Now imagine the difficulties of getting thousands of individuals you've never met to embrace change. Even a shining new iPhone doesn't guarantee a sweeping "Huzzah!" across the organization. Internal adoption of your BDAs won't—and *shouldn't*—happen overnight. You'll need a proven plan of attack.

To illustrate the right plan, let's revisit the story of our old friend Ruairi.

RUAIRI'S PLAN FOR PIECEMEAL CHANGE

When last we left Ruairi, he had engaged an outside development firm to create a mobile portal for Infield, the company's project management platform for foremen.

Even after getting the go-ahead, Ruairi expects a continual learning curve. "Explaining my rationale across the organization has been a challenge," he says. "Everyone has to understand why it's important. Everyone has to understand why it helps them."

As a platform, Infield is low-hanging fruit, ripe for mobility. Ruairi's plan for a more accessible technological infrastructure is to whet worker appetites and roll out more apps as the organization gets hungrier. He calls this the "chicken nugget" approach. "You can't feed a chicken dinner to people who aren't hungry or aren't sure they like chicken," Ruairi explains. "You put it in front of them and they'll push it away, telling you they aren't hungry. But give them a chicken nugget at a time, and they'll take more until they've finished the meal."

Oldcastle foremen who had never owned PCs until the introduction of Infield are now among the most vocal when it comes to technology decisions. Piecemeal change within the organization is just simpler for entire departments to digest.

Oldcastle's big picture is Ruairi's "chicken dinner:" he hopes to create a ripple effect that will extend into more valuable apps for on-site workers. That's only the tip of the iceberg, though. He's also focused on bringing accessibility to other roles, both from within and outside of the company—from sales professionals to partner vendors.

Ruairi's plan highlights an important element of the post development process: a continuously improving technological infrastructure. It's a point worth further discussion.

CHANGE MANAGEMENT AND YOU

Throughout the book, we've demonstrated how enterprise mobility is progressive. Organizations are made up of humans, and humans are resistant to change. Smaller changes, stretched across scheduled timeframes, are easiest to digest for the end users within your organization.

Therefore, change management cannot be taken lightly. A great mobility program builds excitement around the products, ensuring that your workforce embraces change.

A.M. Castle and Co. gets it. In their first major mobile push, the organization unwrapped a program meant to build hype around the new apps. They even created videos to demonstrate how much faster and easier the new tools made the process they addressed. The company's rollout campaign included a concerted effort to compare the new system with the old. The previous system had earned the nickname "Hourglass" for how slow it moved. Internal app champions branded the new one "Cheetah" for its speed, going so far as to create T-shirts touting the new system.

Behind the scenes, A.M. Castle and Co. was able to actively track usage on a daily basis. These numbers helped them understand how quickly users were migrating. Visibility into adoption played a huge role in ensuring a smooth rollout.

Sonic Automotive gets workers excited about mobility too. The company's dedicated change management program keeps everyone engaged in Sonic's continuously evolving technological infrastructure.

Rather than leaving app end users out in the cold, company experts take to the road for months to brief members of the workforce at more than a hundred physical locations throughout the country. Sonic's "Carpe Diem" off-site meetings with associates usually last two to four hours. Members of the workforce get a peek into the organization's technological vision for the year and their roles in the evolving tech landscape.

Like A.M. Castle and Co., Sonic configured sensors to monitor the health of user adoption. They were able to determine which dealerships were fully engaged and which needed additional help to get everyone on board.

End users aren't the only people who need change management. Shiny object syndrome—pushing for new tech developments the moment a fresh tool, strategy, or idea appears on the market—used to be a problem for Sonic. Through change management, the organization now takes a methodical approach to app development, slowing down the process to ensure it's choosing the right app and engaging staff members in the rollout process.

As a change agent, it's natural to want to bring your organization up to speed as quickly as possible. Just getting executive buy-in suddenly lays the groundwork for any number of mobility projects, some of which could even put you way ahead of the competition.

Enterprise organizations are big, complex beasts. Mobility therefore carries the cargo of iterative change. Continuous development is the product of strong change management. Understanding the importance of pacing enterprise mobility is a huge advantage. It's how you ensure that you get workforce buy-in for each step in a continuous process toward a more accessible enterprise.

Ruairi's "chicken nugget" method represents the right course for Billion Dollar Apps. You can't do everything at the same time. Do a little the first time, and do it correctly. That way, you'll open up the door to more changes down the road.

Future changes that result from your initial mobility project may include:
- New mobile apps and devices for other departments
- More functionality for the initial suite of apps

- Stronger integrations between platforms and systems that should work together (but often don't)
- Workflow revision across the enterprise that makes life easier for workers across many different roles

In essence, your initial BDA project's success is more than just its own impact on the organization. That success is also based on what the project does for the organization's technology core as a whole. You're setting the tone for a decade of technological prosperity. That's not just a victory—it's a legacy. Of course, you have to start somewhere.

GROOMING CHAMPIONS FOR TECHNOLOGICAL CHANGE

As a change agent, you understand the time and effort that goes into enterprise-wide change. By the end of the BDA process, you'll have an even more intimate understanding of it.

The best thing you can do for your organization's new technology is to pass your desire along. Others need the understanding that this new technology is not only good for the organization—it's good for each and every person who gets to use it on a daily basis.

Grooming champions for your new technology isn't as hard as you might think. The testing process is multifaceted in that it's also an opportunity to impart ownership on a group of end users. Leaders within your workforce will naturally invest in your apps when you give them a say in the final product.

That investment can pay huge dividends in platform adoption across the rest of your workforce. A few influential workers getting on board with the new apps can set the trend for the rest of the workforce, helping to start the ripple effect that leads to a mobilized workforce. One day soon, accessibility can become a concrete thing for the entirety of your organization.

Your app champions are a major part of the equation. You should tackle education on an enterprise-wide scale through a compelling training program.

SMOOTH APP ADOPTION THROUGH TRAINING AND VISIBILITY

One point of introducing mobility into the enterprise is to limit the resources you need for training. Yes, Billion Dollar Apps are meant to be easy enough to figure out on the user's own. But they won't be figured out at all if your end users don't play with the new apps, let alone even open or download the suite.

Rolling out mobility requires a coordinated training program. It doesn't have to be extensive. You won't even have to continuously train new employees on it after you get your workforce on board, in most cases.

The point behind your training program is to introduce your workforce to the new technology and to walk staff members through how to use it. The training is simply an acclimation program. It helps with app visibility. Most important, training is about communicating the value of your apps to the end users so they understand the benefit to them.

A strong training program may consist of the following elements:

- An e-mail or other department-wide communication announcing the new technology, sent a month prior to debut
- A department meeting, conference call, or webinar organized by your marketing director and presided over by regional managers
- A series of short videos explaining how to use the new technology, what to do with broken devices, etc.
- A handful of IT professionals or managers familiar with the software, dispatched to locations to help personnel get hands-on with the new technology

Training works differently depending on your organization. The devices and apps should be intuitive, so you won't require long training sessions over several weeks. All you need is a bit of time to get your workforce acquainted with your new technology.

Your training program shouldn't just address the end users, either. Your help desk team should get training on the app, but also guidance on the basics of mobility. People too often have app issues that are actually related to devices and connections—the difference between connecting via cell signal or the in-office corporate Wi-Fi, for instance.

The battle for a great mobile experience doesn't end after you get workforce buy-in. You must build a continuous support structure for your apps.

PHASE 6: SUPPORT

Mobility requires upkeep. Without a nurturing hand, your apps will quickly become obsolete.

Support and ongoing maintenance are critical to the post development phase of mobility. You have initial momentum, but you need to keep the ball rolling. The best practices we've covered already should guide you as you produce new apps and make updates to old ones. Needs, priorities, and available resources should govern how often you create new apps. Your development team should plan to push updates every one to three months. Updates should take high priority bugs, user-sourced improvements, and new device or operating system requirements into account.

Here are three tips to help define your app maintenance strategy as you go about updating your mobility program.

Source update ideas from the field.

Your workforce will have ideas on how mobility can better help them accomplish their jobs. Embrace feedback and ideas from the field in future updates to ensure a steadily improving mobile experience for your organization. Create a continuous feedback loop where people can submit feedback as they use your apps.

Rely on your mobile champions.

Remember, you should be maintaining a corps of mobile champions. Their proximity to the mobile experience and mobility's organizational goals put them in a unique position to help you improve your apps.

Watch the numbers.

With instrumentation in place, you have the ability to measure how people interact with your apps, where apps appear to be lacking, and whether or not your backend system is handling communication well. Analyze those numbers and use them to understand improvements that even the people using your apps might not have considered.

When all is said and done, your mobile program only stretches as far as the support structure you've built around it. Keep mobility at the forefront of your tech planning, and you'll set your enterprise mobile program up for success.

DISCOVERING YOUR BILLION DOLLAR APPS

The title of this book may be catchy, but "app" is a loaded term. As consumers, we've come to think of apps as those fun, sometimes useful tools we download to our smartphones and tablets.

But, as you now know, enterprise mobility is about much, much more than devices and software. These are the tools that unlock the benefits of Billion Dollar Apps. Reimagining your business is the heavy lifting. To do it right, you have to be ready to ask the question, "What if?" You have to be ready to ask it over and over again...and you can't be skeptical when the answer comes back: "This is something you can do right now on your budget." The truth is that any enterprise organization under the sun has millions of possibilities within its daily operations to make things easier for its workforce. Each of these opportunities is worth thousands, millions, even *billions* of dollars to your organization over the long run.

Billion Dollar Apps aren't manufactured from scratch. They're running a low profile just below the surface of your enterprise, waiting to be discovered by the first change agent to come along and make a few waves.

It's not an easy process. But I hope the tools in this book can help you make sense of it. The time to go mobile is now. Why wait? Get out there and discover your organization's Billion Dollar Apps.

CHAPTER 13
Taking the Plunge

The pace of mobile app development continues to accelerate. Still, very few organizations are treating mobility as strategic. They fixate on the obvious apps and limit their own potential.

I wrote this book to share what I've learned through working with enterprises to achieve billion-dollar results with apps. I've worked with a wide range of organizations, each with a unique business model and fresh ideas on how to improve it. It's thrilling to see these ideas come to life in incredible and diverse ways every day.

We've helped companies much like Protectus revamp their sales processes. We've put life-changing mobile tools in the hands of hundreds of thousands of workers. We've even worked on a rocket launcher control system (so, for those of you familiar with the *Terminator* series: I'm sorry, but we helped build Skynet).

I share the lessons of my experiences here, but I can't share the experiences themselves. You'll have to earn some on your own. My purpose is to get change agents like you into a mobile mindset. The hands-on advice, equations, and strategy explanations are meant to give you structure for the part of the journey you can undertake with your colleagues.

Going it alone in app development can be an arduous process filled with starts, stops, and dead ends. The information in this book is meant to get you past as many of these as possible, ensuring a smooth process that kick-starts a rich future of mobile development.

Striking out into the icy waters of technological change unaided works only for the most rugged explorers—those change agents with deep experience in implementing mobility. The rest of us would do well to come equipped with a navigator and a life vest. As such, you'll need guidance along the way.

I hope I have started a conversation. You'll have questions, triumphs, pain points, insights, and regrets to contribute to it. I encourage you to reach out and share. I'm always looking for new stories to tell, of either successes or failures, that we can all learn from.

Reach out to me:
- E-mail: alexb@lextech.com
- Phone: 1-630-420-9670
- LinkedIn: http://www.linkedin.com/in/alexbratton
- Twitter: https://twitter.com/alexbratton

If you're curious to learn more about Lextech, check out our website: www.lextech.com. I wish you the best of luck planning out the Billion Dollar Apps for your organization.